# International Cryogenics Monograph Series

The International Eryogenics Monograph Series was established in the early 1960s to present an opportunity for active researchers in various areas associated with cryogenic engineering to cover their area of expertise by thoroughly covering its past development and its present status. These high level reviews assist young researchers to initiate research programs of their own in these key areas of cryogenic engineering without an extensive search of literature.

More information about this series at http://www.springer.com/series/6086

J.G. Weisend II
Editor

# Cryostat Design

## Case Studies, Principles and Engineering

Springer

*Editor*
J.G. Weisend II
European Spallation Source (ESS), ERIC
Lund University
Lund
Sweden

ISSN 0538-7051 ISSN 2199-3084 (electronic)
International Cryogenics Monograph Series
ISBN 978-3-319-80976-2 ISBN 978-3-319-31150-0 (eBook)
DOI 10.1007/978-3-319-31150-0


Softcover reprint of the hardcover 1st edition 2016


Printed on acid-free paper

This Springer imprint is published by Springer Nature
The registered company is Springer International Publishing AG Switzerland

*For Shari, Rachel, Alex*
*and Nick—As Always*

# Preface

Cryostats are technical systems that maintain equipment or cryogenic liquids at cryogenic temperatures. As such, they are one of the fundamental building blocks of cryogenic systems. Examples of cryostats include the magnet cryostats that comprise the majority of the Large Hadron Collider (LHC) at CERN, spaceborne cryostats containing sensors operating below 1 K, MRI cryostats found in most large hospitals, and large cryogenic liquid storage vessels. Cryostats that contain superconducting radio frequency cavities are frequently referred to as cryomodules, while cryostats whose principal function is to store cryogenic fluids are also referred to as dewars. Cryomodules and dewars are also covered in this work.

The proper design of cryostats requires the knowledge of many disciplines including cryogenic properties of materials, heat transfer and thermal insulation, instrumentation, safety, structures, and seals. One of the best ways to learn about cryostat design is to study the design choices and resulting performance of previous designs.

This book provides such a review. It starts with an introductory chapter on the principles of cryostat design including practical data and equations. This chapter is followed by a series of case studies on existing cryostats. The studies describe the cryostat and the design choices made along with the resulting performance of the cryostat. The cryostat examples used in the studies are chosen to cover the wide range of cryostat applications and the authors of each case are leading experts in the field, all of whom participated in the design of the cryostats being described.

Chapters 2 and 3 are case studies involving superconducting magnets for large particle accelerators. Due to the large numbers of magnets required in these cases, low heat leak, reliability, and cost are key requirements. Chapter 4 describes a one of a kind spaceborne dewar system whose requirements are very different than that of accelerator cryostats. Chapters 5 and 6 describe cryomodules that contain superconducting RF cavities in particle accelerators. As will be seen there are two broad families of these cryomodules (segmented and continuous) with different design drivers and approaches. Taken together these chapters describe a total of six different cryomodules and the evolution of cryomodule design from the 1980s to the 2010s can be seen. Chapter 7 presents special topics in cryostat design. These topics are of particular importance for MRI magnet cryostats; which provide the examples but are broadly valuable for all cryostats. A cryostat design for very low (50 mK) temperatures is described in Chap. 8. In addition to the lower temperatures, this cryostat has unique material requirements due to the need to keep the radioactive background of the associated experiment as low as possible. Transfer lines connect cryostats and are a type of cryostat themselves. Transfer line features, an overview of major transfer line systems, and a detailed case study of a transfer line are found in Chap. 9. The final chapter provides a summary by listing guidelines for a successful cryostat design. Extensive references throughout provide sources of further information.

Lund, Sweden J.G. Weisend II
June 2016

# Acknowledgments

This book would not have been possible without the contributions from the chapter authors, all of whom are involved in ongoing projects with very demanding schedules. I wish to express my thanks to my co-authors for their hard work. Tom Spicer, Cindy Zitter, and the team at Springer have been very supportive and helpful in the development and production of this book. I am honored that this book is being published as part of the long-running *International Cryogenic Monograph Series*. Professor Steven Van Sciver, in addition to being one of the series editors, was also my thesis advisor at Wisconsin and I would like to thank him for making this book possible in many ways than one. My colleagues at the European Spallation Source and Lund University have also been a great help.

As always, the help and support of my family has been the key to the completion of this book. For Shari, Rachel, Alex, and Nick—my continuing gratitude.

Lund, Sweden
June 2016

J.G. Weisend II

# Contents

# Contributors


**E. Daly** Thomas Jefferson National Laboratory, Newport News, VA, USA

**Michael DiPirro** Code 552, NASA Goddard Space Flight Center, Greenbelt, MD, USA

**Jaroslaw Fydrych** European Spallation Source ERIC, Lund, Sweden

**Philippe Lebrun** CERN, European Organization for Nuclear Research, Geneva, Switzerland

**Thomas H. Nicol** SRF Development Department, Technical Division, Fermi National Accelerator Laboratory, Batavia, IL, USA

**T.J. Peterson** Cryomodule Engineering, SRF Development Department, Technical Division, Fermi National Accelerator Laboratory, Batavia, IL, USA

**J. Preble** Thomas Jefferson National Laboratory, Newport News, VA, USA

**Richard L. Schmitt** Fermi National Accelerator Lab, Batavia, IL, USA

**Wolfgang Stautner** GE Global Research, Electromagnetics and Superconductivity Lab, One Research Circle, Niskayuna, NY, USA

**J.G. Weisend II** European Spallation Source ERIC, Lund University, Lund, Sweden

# Chapter 1
# Principles of Cryostat Design

**J.G. Weisend II**

**Abstract** This chapter provides an introduction to the engineering and design of cryostats. It reviews cryostat requirements and provides detailed technical information on topics relevant to cryostats. These topics include: materials, heat transfer and thermal insulation systems, structural supports, safety, instrumentation, seals and connections, transfer lines and thermoacoustic oscillations. The role of prototyping and series testing in cryostat development is also discussed. Numerous tables, figures and equations provide useful information for the cryostat designer.

## 1.1 Cryostat Requirements

Successful engineering design depends upon a thorough understanding of the requirements of the component being designed. Rushing into design before fully understanding the requirements can lead to poorly optimized designs and a loss of time and resources.

Cryostats typically have a set of common requirements, not all of which can be simultaneously met to the same degree. Explicitly determining the requirements of a cryostat and determining which ones are most important is the key first step to a successful cryostat design. The following are requirements that should be considered in cryostat design.

1. Operating Temperatures and Allowable Heat Load

Since the purpose of a cryostat is to maintain systems at cryogenic temperatures, the definition of the operating temperatures and the allowed heat load at those temperatures are generally some of the first requirements defined. These requirements drive a number of aspects of the resulting design including: material selection, design of the thermal insulation system, cost, complexity and size. There are a

J.G. Weisend II (✉)
European Spallation Source ERIC, Lund University,
P.O. Box 176, 22100 Lund, Sweden
e-mail: john.weisend@esss.se

 
J.G. Weisend II (ed.), *Cryostat Design*, International Cryogenics Monograph Series,
DOI 10.1007/978-3-319-31150-0_1

number of subtleties in these requirements. An important distinction is that between static heat leak and dynamic heat load. Static heat leak refers to heat entering the cryostat from the outside environment. In many cryostats, this is the sole or principal source of heat deposited at cryogenic temperatures. However, in other cryostats there may be an additional heat load to cryogenic temperatures that stems from the operation of the cryogenic equipment itself. Such heat loads typically vary with time and are thus dynamic. Examples of such heat loads include heating of the cavity walls in superconducting radiofrequency systems and heat deposited by particle beams in accelerator or target cryostats. Dynamic heat loads can be the dominant heat loads to the cryostat, which may affect the design choices made in the cryostat's thermal insulation system.

The temperatures at which heat is deposited in the cryostat are determined by a number of factors. The required operating temperature of the cryogenic equipment is one requirement but temperatures of intermediate shields or heat sinks may be driven by the optimal performance of small cryocoolers or large cryogenic refrigeration plants.

Cryostats are frequently part of a larger cryogenic system. Allowable heat loads and operating temperatures must be consistent with the larger system needs and overall system optimization may be possible [1]. The allowable heat loads in a cryostat may also be related to the number of cryostats in the system. A system with a large number of cryostats, such as LHC or International Linear Collider (ILC) may require a smaller heat load per cryostat to keep the overall cryogenic system a reasonable size. Such systems put additional effort into reducing the heat load of each cryostat (see case studies in the following chapters).

2. Alignment and Vibration

Depending on the application, some cryostats have requirements on the alignment of the cryogenic equipment they contain or have limits on their allowable vibrations. Issues to consider when responding to these requirements are the impact of repeated thermal cycling on the alignment and the need, if any, to adjust the alignment while the cryostat is cold or under vacuum Alignment requirements can be quite tight. In the case of the TESLA cryomodules [2] the magnets had to maintain an alignment to $\pm 0.25$ mm relative to an ideal reference.

3. Safety

Safety requirements including maximum credible accidents should be determined at the start of the design and any design implications factored in from the beginning. Retrofitting a design to meet safety requirements can be very expensive and time consuming. In particular, issues due to overpressure and venting as a result of vacuum or pipe failures can require significant design effort [3].

4. Allowable Size and Weight

These are particularly important for space borne cryostats but are also relevant for cryostats that must be transported.

5. Amount and Type of Required Instrumentation

This may well vary between prototype and mass produced cryostats.

6. Number of Feedthroughs for Instrumentation, Power, External Manipulators, Optical Windows and Cryogenic Lines.
7. Ease of Access to Cryostat Components
8. Existence of Ionizing Radiation or Magnetic Fields

This will influence material and instrumentation choices. The performance of superconducting RF cavities, for example, is strongly affected by local magnetic fields. This results in a need for supplemental magnetic shielding in the cryostat as well as a limit on magnetic material near the cavities.

9. Additional Material Requirements

Some cryostats have additional unique material requirements. For example the Cryogenic Dark Matter Search cryostat (Chap. 8) requires the use of materials with no radioactivity to avoid false signals. This severely constrains the material choices.

10. Existing Regulatory Code Requirements

These might include pressure vessels codes or codes associated with flammable materials.

11. Expected Lifetime of the Cryostat
12. Number of Cryostats Required

Is this a single unique cryostat or will it be mass-produced? Different design choices may be made in either case.

13. Cost and Schedule Limits

These should be known and considered at the beginning of the design.

All designs involve compromise and it's extremely unlikely that all cryostat requirements can be met equally. The trick is to properly prioritize the requirements in order to create an optimal design. Figure 1.1 shows an example of this approach. The figure shows the E158 liquid hydrogen target cryostat operated at the SLAC National Accelerator Lab [4]. This cryostat contained a loop of circulating $LH_2$ operating at 20 K. A high energy electron beam was scattered by the $LH_2$ as part of a fundamental physics experiment. The presence of the beam had two important effects: the cryostat was exposed to a extremely high ionizing radiation environment and the beam deposited roughly 900 W of heat into the 20 K $LH_2$. This resulted in two important design choices. First, materials such as kapton based multilayer insulation and ceramic based electrical insulation that survive high radiation doses were chosen for the cryostat. Second, given the high dynamic heat loads from the beam, expensive techniques to reduce the static heat load, such as actively cooled thermal radiation shields were not used.

**Fig. 1.1** View of the upper portion of the SLAC E158 liquid hydrogen target cryostat [4]

## 1.2 Cryogenic Properties of Materials

Selection of the proper materials is an important aspect of cryostat design. Many materials are unsuitable for use at cryogenic temperatures and the properties of materials can change greatly between room temperature and cryogenic temperatures. Failures in cryostats can frequently be traced to the use of inappropriate materials or to not accounting for the change in material properties as a function of temperature.

The following materials are suitable for use at cryogenic temperatures:

- Austenitic stainless steels e.g. 304, 304L, 316, 321
- Aluminum alloys e.g. 6061, 6063, 1100
- Copper e.g. OFHC, ETP and phosphorous deoxidized
- Brass
- Fiber reinforced plastics such as G-10 and G-11
- Teflon (depending on the application)
- Niobium and Titanium (frequently used in superconducting RF systems)
- Invar (Ni/Fe alloy) useful in making washers due to its lower coefficient of expansion
- Indium (used as an O ring material)
- Kapton and Mylar (used in Multilayer Insulation and as electrical insulation
- Quartz (used in windows)

Some materials unsuitable for use at cryogenic temperatures include:

- Martensitic stainless steels—undergoes ductile to brittle transition when cooled down.
- Cast Iron—also becomes brittle
- Carbon steels—also becomes brittle. Sometimes used in 300 K vacuum vessels but care must be taken that breaks in cryogenic lines do not cause the vacuum vessels to cool down and fail.
- Rubber and most plastics.

One should never use materials in cryostats unless their suitability for use at cryogenic temperatures is verified and their behavior as a function of temperature is well understood. Given the large, and frequently nonlinear, changes in material properties as a function of temperature, simple extrapolation of room temperature properties to cryogenic temperatures should never be done. References [5–9] are good starting points for finding the cryogenic properties of materials. Searching in the proceedings of the International Cryogenics Materials Conference series may also be useful. There are also commercially available computer codes that produce material properties at cryogenic temperature [10, 11]. Aspects of some key cryogenic properties of materials are given below.

### 1.2.1 Thermal Contraction

Almost all materials change size (most shrink) when cooled to cryogenic temperatures. This effect can be significant and can cause a number of issues in cryostats including the development of interferences or gaps between adjacent parts upon cooling, adverse impact on alignment and possible failure upon cool down in over constrained components or wires.

Thermal contraction should always be considered in cryostat design. Guidelines for design include:

- Allow components room to contract through the use of sliding joints, bellows or only fixing at a single point
- Install sufficient slack into wire and cable assemblies to allow for contraction
- Overlap thermal shields and multilayer insulation to avoid gaps upon cool down and contraction
- Analyze the effect of thermal contraction on system alignment and on the possibility of interferences between developing between adjacent components
- Play particular attention to the impact of thermal contraction on different materials connected to or adjacent to each other

Thermal contraction is generally described by the property of thermal expansivity [12] which is the change in volume or length of a material with temperature. This property, which itself is a function of temperature, is of limited use in cryostat

**Table 1.1** Integrated thermal contraction for practical engineering materials between 300 and 100 K and between 100 and 4 K

| Material | ΔL/L (300–100 K) | ΔL/L (100–4 K) |
|---|---|---|
| Stainless steel | $296 \times 10^{-5}$ | $35 \times 10^{-5}$ |
| Copper | $326 \times 10^{-5}$ | $44 \times 10^{-5}$ |
| Aluminum | $415 \times 10^{-5}$ | $47 \times 10^{-5}$ |
| Iron | $198 \times 10^{-5}$ | $18 \times 10^{-5}$ |
| Invar | $40 \times 10^{-5}$ | – |
| Brass | $340 \times 10^{-5}$ | $57 \times 10^{-5}$ |
| Epoxy/fiberglass | $279 \times 10^{-5}$ | $47 \times 10^{-5}$ |
| Titanium | $134 \times 10^{-5}$ | $17 \times 10^{-5}$ |

design. More useful is the integrated change in length of a material between two end temperatures. This is shown in Table 1.1 for a number of practical engineering materials.

In order to find the change in length per unit length from 300 to 4 K, simply add the two columns together. Thus, on average, a 1 m length of stainless steel will contract $331 \times 10^{-5}$ m or 3.3 mm. Notice that Invar, a Iron–Nickel alloy contracts very little as a function of temperature. Invar has been used in cryostats (see Chap. 5) where very little length contraction is desired.

### 1.2.2 *Thermal Conductivity*

The thermal conductivity of materials, particularly metals, varies strongly with temperature between 300 and 0 K. This variability must always be taken into account in cryostat design. Figure 1.2 shows the thermal conductivity of a number of materials as a function of temperature. In many cases, the analysis may be simplified by the use of thermal conductivity integrals [8]. In this approach, the conduction heat transfer in one dimension is given by:

$$Q = -G(\theta_1 - \theta_2) \tag{1.1}$$

where Q is the total heat transferred by conduction and the Geometry Factor (G) is given by:

$$G = \frac{1}{\int_{x_1}^{x_2} \frac{dx}{A(x)}} \tag{1.2}$$

Here x is the position along the conduction path and A is the cross sectional area. Note that in the case of a component with length L and constant cross sectional area A, Eq. 1.2 becomes simply A/L.

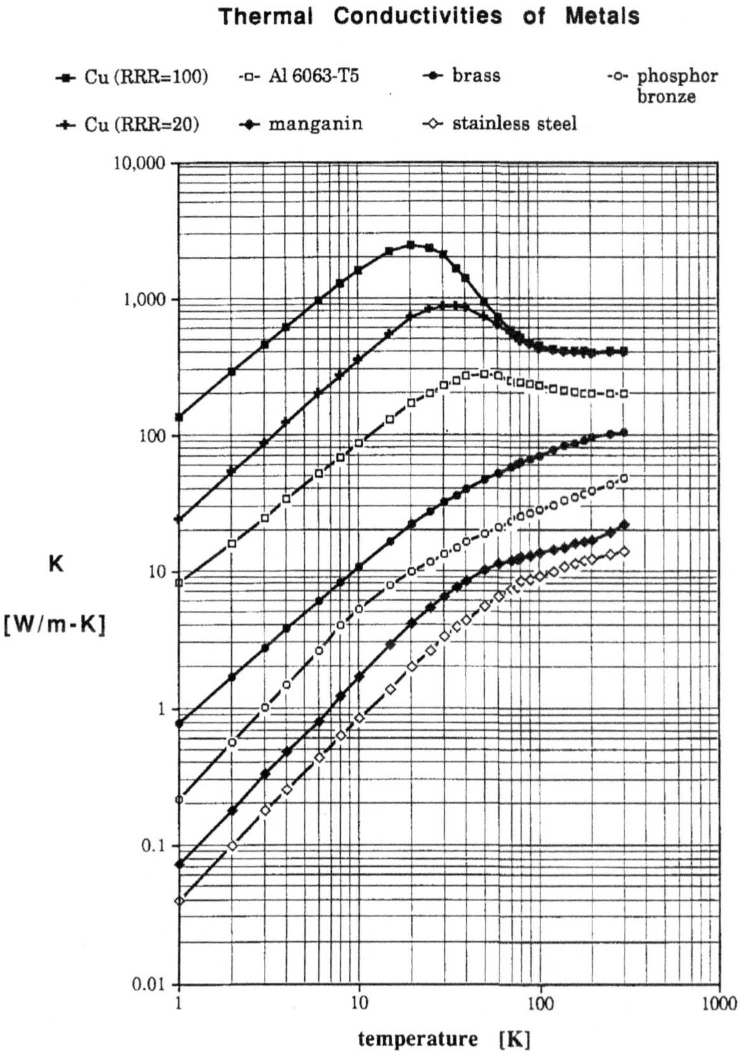


**Fig. 1.2** Thermal conductivity as a function of temperature for a variety of engineering materials [8]

The thermal conductivity integral $\theta_\iota$ is given by:

$$\theta_i = \int_0^{T_i} k(T)dT \tag{1.3}$$

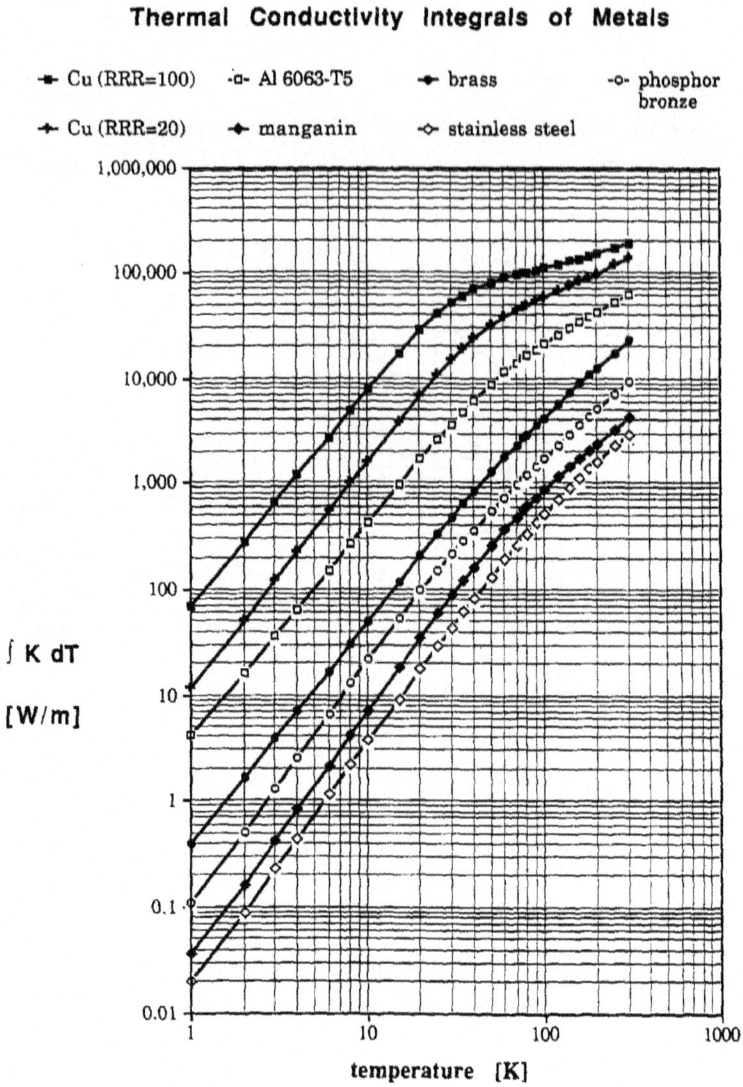


**Fig. 1.3** Thermal conductivity integrals of a variety of engineering materials [8]

where T is the temperature and k is the thermal conductivity. Knowing the appropriate thermal conductivity integral for a given material and temperature allows quick calculation of the heat transfer between known temperatures while still taking into account the variation of thermal conductivity with temperature. Figures 1.3 and 1.4 show thermal conductivity integrals for a range of materials.

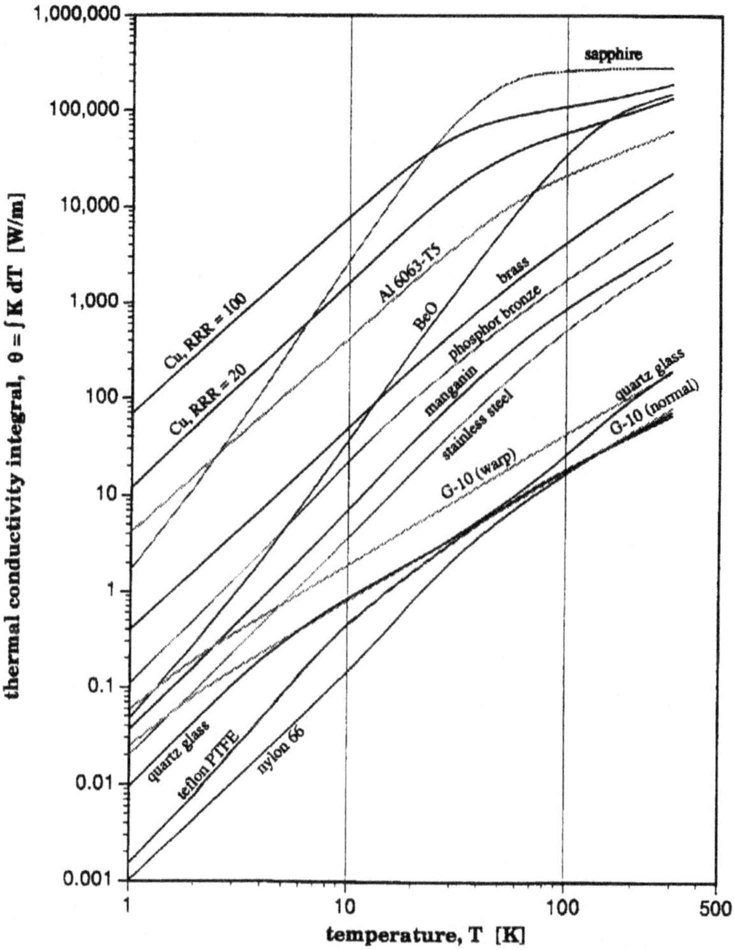


**Fig. 1.4** Thermal conductivity integrals of a variety of engineering materials—Courtesy of Lakeshore Cryotronics

## *1.2.3 Heat Capacity*

The heat capacity or specific heat of materials generally decreases as a function of temperature. This has two principal effects in cryostats. First, for a constant rate of cooling, the drop in temperature over time increases as the cryostat becomes colder. This effect is easily seen in larger systems which cool down more rapidly at lower temperatures. Second, at cryogenic temperatures, the lower heat capacity means that small heat leaks can cause significant temperature rises. This can be particularly important for system operating below 1 K. Figure 1.5 shows the specific heat as a function of temperature for a number of engineering materials.

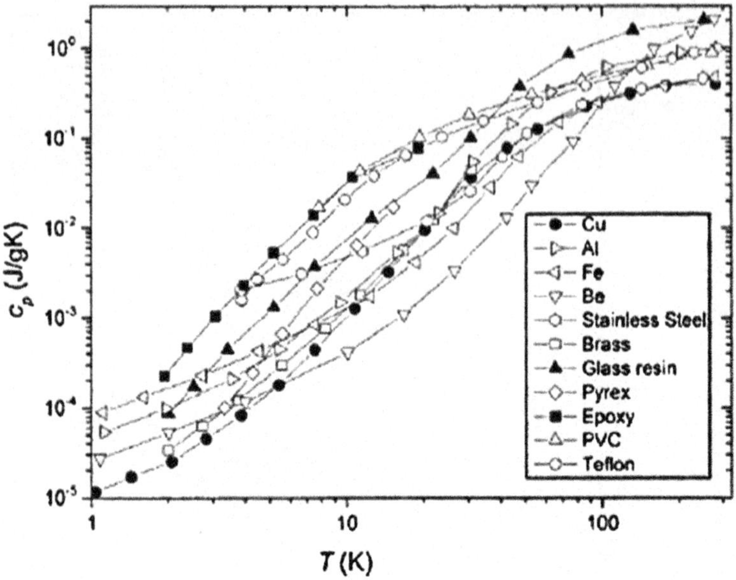


**Fig. 1.5** Specific heat as a function of temperature for a variety of engineering materials [9]

### *1.2.4 Material Strength*

Material strength properties such as yield strength and ultimate tensile strength also change with temperature. In many metals, those that don't become brittle, the strength actually increases with decreasing temperatures. However in this case, most cryostats are designed using the lower, more conservative values at 300 K. The reason for this is two-fold. First, all cryostats start out at room temperature and second, due to operational issues even a cold cryostat may return suddenly to room temperature.

There are some cases in which the higher values at lower temperatures may be taken into account. In the case of Superconducting RF cavities made from niobium, the increased strength of the niobium at cryogenic temperatures may be used when analyzing the response of the cavity to accidents that only occur when the cavity is at cryogenic temperatures. An example of such an accident is the loss of beam tube vacuum and the subsequent condensation of air on the surface of the cavity. This condensation deposits heat and rapidly boils away the helium surrounding the cavity causing large pressure rises in the helium space. Such a scenario can't occur if the cavity is at room temperature and thus the cavity's stronger, lower temperature strength properties can be used in determining the cavity's response to the sudden pressure rise.

As always, the material strength properties should always be checked at the temperatures of interest to the cryostat being designed.

## 1.3 Thermal Insulation and Heat Transfer

A primary function of a cryostat, of course, is to maintain the equipment inside at cryogenic temperatures. This is accomplished by developing thermal insulation systems that block heat from the outside environment from reaching the cryogenic equipment. While a perfect insulation system doesn't exist, properly designed cryostats can reduce the heat leak to very low, even μW, levels. This is accomplished by interrupting the three mechanisms of heat transfer within the cryostat: conduction, convention and radiation.

### *1.3.1 Reducing Conduction Heat Transfer*

The Fourier heat conduction equation (Eq. 1.1) shows that the heat transferred by conduction is proportional to the temperature difference, the thermal conductivity and the cross sectional area of the conduction path and is inversely proportional to the length of the conduction path. Thus, the conduction heat leak can be reduced by using low thermal conductivity materials, reducing the cross sectional area of the conduction path and increasing the length of the conduction path. As you read through the case studies later in this book, you will notice again and again, the presence of long, thin conduction paths made from low conductivity material. Another way to reduce the conduction heat leak to the lowest temperature is to intercept the heat at an intermediate temperature. This reduces the temperature difference and thus the heat to the lowest temperature point. The heat conducted from higher temperatures to the intermediate heat sink still has to be removed but it is thermodynamically more efficient to remove the heat at a higher temperature. Chapter 7 gives an example of the optimization of such heat sinks.

Figure 1.6 shows the support post first developed for the SSC magnet cryostat and shown here for use on the TESLA (now ILC) cryomodule [13]. The design of this post reduces the conduction heat leak via the principles described above. The conduction path in the post is made from low conductivity G-10 cylinders with a small cross sectional area. In addition, active heat sinks are installed at the 77 K and 4 K levels. As a result, the post transfers only 0.03 W of heat between room temperature and 2 K over a distance of 140 mm while still being able to support a load of 50 kN.

### *1.3.2 Reducing Convection Heat Transfer*

Convection heat transfer is that caused by a moving fluid. In many cryostats, this mode of heat transfer is eliminated by using a vacuum space to separate the cold and warm portions of the cryostat. This removes any fluid that can transfer heat via

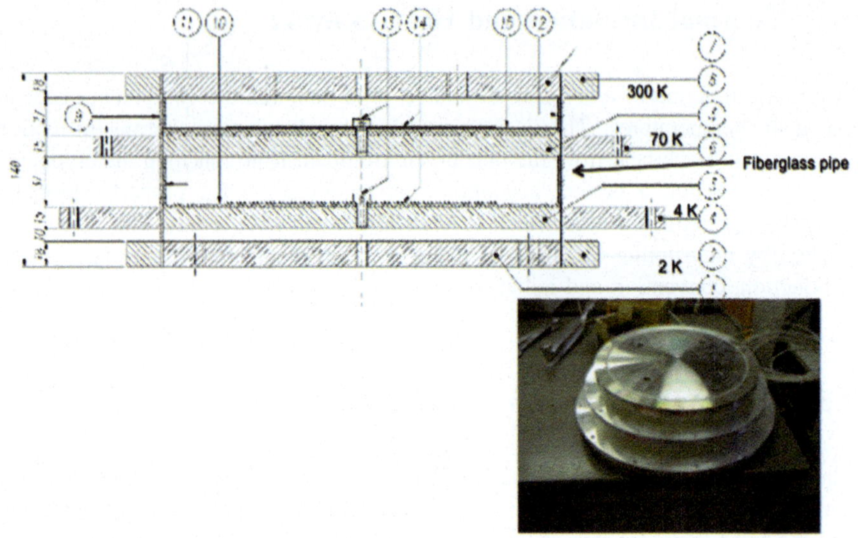


**Fig. 1.6** The support post for the ILC cryomodule—Courtesy of Nicol [13]

convection. The use of such an isolation vacuum was a key feature of James Dewar's original liquid hydrogen cryostat.

In general, a pressure of $10^{-6}$ millibar or less is sufficient vacuum to eliminate all convective heat transfer. The cold wall of the vacuum space acts as a cryopump for residual gases. Cryostats, particularly those operating below 77 K may have isolation vacuums of $10^{-8}$ millibar or less when cold. It is not good practice to depend upon cryopumping only but rather to use mechanical pumps to reduce the pressure to near $10^{-6}$ millibar prior to cooling the cryostat down. In some cases, the use of isolation vacuum spaces is not possible or practical and other techniques such as foam insulation are used; see Sect. 1.3.4.

### *1.3.3 Reducing Radiation Heat Transfer*

All surfaces emit heat via electromagnetic radiation. This thermal radiation is frequently the largest source of heat leak into a cryostat. There are a number of techniques for reducing the radiation heat leak into a cryostat. These techniques are driven by the applicable governing equation for idealized (infinite parallel plates, surface emissivities equal and $\ll 1$) radiation heat transfer [8]:

$$q = \frac{\epsilon}{(N+1)*2}\sigma\left(T_H^4 - T_L^4\right) \tag{1.4}$$

where q is the heat transferred in $W/m^2$, ε is the emissivity of the surfaces, σ is the Stefan-Boltzmann constant ($5.67 \times 10^{-8}$ $W/(K^4m^2)$) and N is the number of uncooled radiation shields between the high temperature surface ($T_H$) and the low temperature surface ($T_L$). Equation 1.4 can also be shown to apply for the case of long concentric cylinders where the gap between the cylinders is small compared to the diameter of the inner cylinder [14]. Based on Eq. 1.4, the heat leak can be reduced by decreasing the emissivity of the surface, adding uncooled radiation shields between the high temperature and low temperature surfaces and reducing the temperature of the high temperature surface.

The temperature of the high temperature surface can be reduced by inserting a lower temperature actively cooled radiation shield between the room temperature (300 K) surface and the lowest temperature surface. The effect of this can be significant. If we use Eq. 1.4 assuming an emissivity of 0.2 and no uncooled shields (N = 0), the heat leak between a 300 and 4.2 K surface is 46 $W/m^2$. If the warmest surface facing the 4.2 K surface is now 77 K then the heat leak is reduced to 0.2 $W/m^2$. The use of actively cooled thermal radiation shields is very common and is seen in most of the case studies that follow.

Actively cooled shields may cooled by cold vapor boiling off from a cryogenic reservoir, cooling flows from a cryogenic refrigeration plant and conductive cooling from a small cryocooler attached to the shield. In many laboratory dewars, a liquid nitrogen bath serves as the thermal radiation shield for the inner liquid helium space.

Figure 1.7 shows emissivities of typical material used in cryostats. When calculating radiation heat leak take care not to underestimate the value of the emissivity; allow for the possibility of tarnishing or oxidizing of the surface that might result in a higher emissivity. Silver plating a surface or coating the surface with highly reflective tape will help reduce the emissivity but can be costly and is generally done only where the absolute minimum radiation heat leak is desired [15] or where there is no room for multilayer insulation.

The reduction of thermal radiation heat leak by the use of uncooled shields to block the line of sight between the cold and warm surfaces can be significant. Multilayer insulation (MLI) also sometimes called superinsulation is the logical extension of this approach. MLI consists of many layers of thin aluminized Mylar or sometimes aluminized Kapton placed into the vacuum space between the warm and cold surfaces. A design feature of MLI is to reduce or eliminate the thermal conduction between the successive layers. This is typically done by placing a nonconductive mesh or paper between the aluminized Mylar sheets or in some cases by only aluminizing only one side of the Mylar and then crinkling the sheets so there are only point contacts between them. Since MLI makes the term N in Eq. 1.4 large, its impact on the thermal radiation heat leak is impressive. As such, most of the case studies that follow employ MLI in their design. Figure 1.8 shows an example of the MLI from the Large Hadron Collider (LHC) magnet cryostat [16].

There are some cautions with the use of MLI. Care much be taken not to pack the MLI too tightly in the vacuum space. Doing so causes the conduction heat load

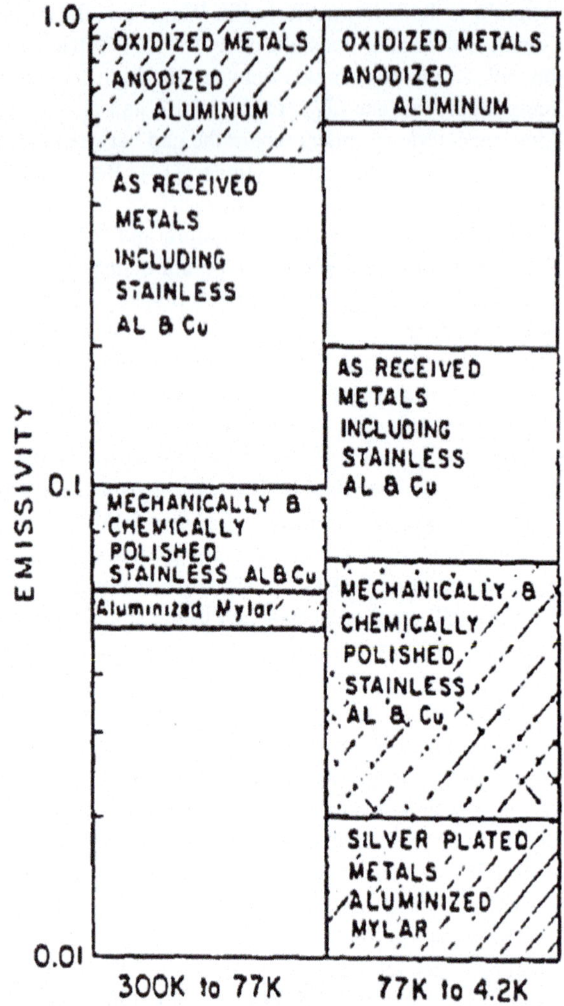


**Fig. 1.7** Emissivities of typical cryostat materials [14]

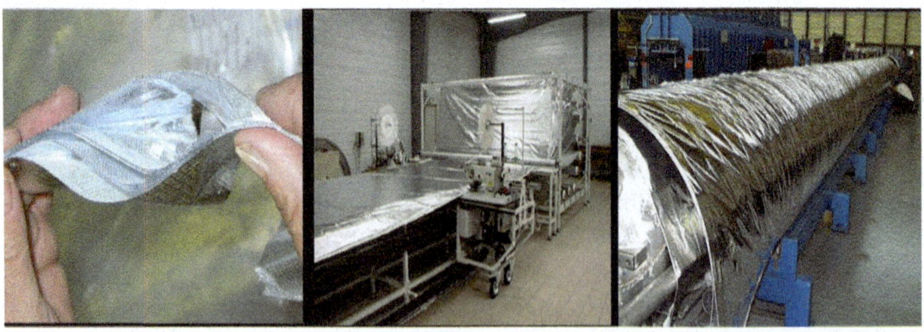

**Fig. 1.8** The MLI system for the large hadron collider (LHC) magnet cryostats. Reproduced with permission from Poncet and Parma [16]. Copyright 2008, AIP Publishing LLC

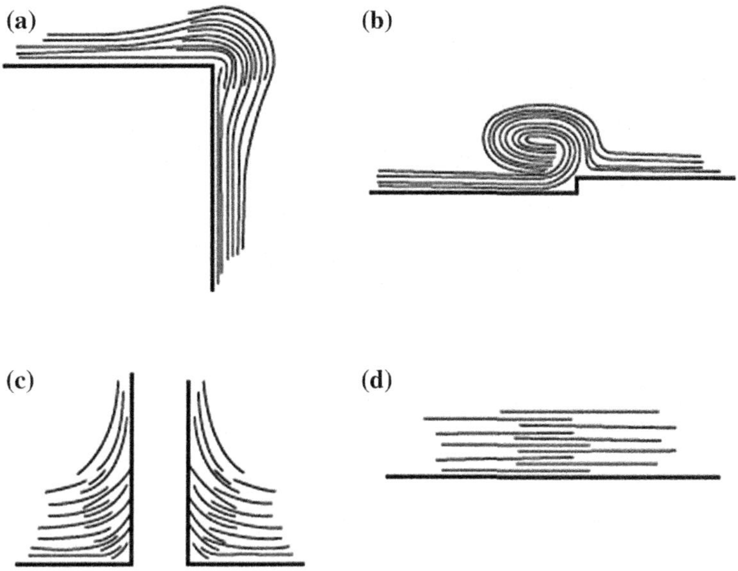


**Fig. 1.9** Examples of the proper installation of MLI for corners (**a**), steps (**b**), penetrations (**c**) and overlaps (**d**) [18]

to increase, reducing the benefit of the MLI. This effect was seen in the SSC Collider Dipole thermal measurement model (see Chap. 2). A good guideline is to keep the density of the MLI in the vacuum space to around 0.5 layers per mm [14]. Another hazard with MLI is associated with seams, ends, penetrations and corners [17]. If these aspects of the installation are not handled properly, their effect can dominate the heat leak. The basic rule here is to avoid connecting the warmest outer layer of the MLI with one of the colder inner layers. This would, in effect, short circuit the MLI and reduces its effectiveness. It is also important to overlap the layers of the MLI at seams so that when the MLI shrinks during cool down, a gap does not open up that allows the warm surface to directly see the cold surface below the MLI. Figure 1.9 illustrates the proper way to address these issues. Detailed analysis and measurements have been made of specific MLI designs and additional information may be found in Refs. [8, 18–20].

### 1.3.4 Other Insulation Approaches

The general approach of using vacuum insulated cryostats combined with actively cooled thermal shields and MLI in the vacuum space is not used in all cryostats. In some cases, such as cryogenic tanks for rocket propellants, the weight penalty caused by the presence of a second wall is too high. In other systems, the cost or complexity of vacuum spaces, MLI and thermal shields out weighs the benefit of a reduced heat leak. This is frequently seen in large storage and transport tanks, particularly for those operating at or above liquid nitrogen temperatures.

One alternative is to remove the vacuum space entirely and coat the cryogenic surface with insulating foam, typically a closed cell foam. This approach was taken with the cryogenic propellant tanks used in the Saturn 5, Space Shuttle and Ariane 5 launchers. This approach is also seen in vent lines and intermittently used liquid nitrogen fill lines mainly to avoid condensing water or air on the cold line surfaces. In the more high performance applications of cryogenic propellant tank insulation, the foam systems can be quite sophisticated, employing water proof barriers and helium gas purging to prevent condensation of water or air. Figure 1.10 shows the foam system for the Saturn 5 s stage liquid hydrogen tank [21]. The overall heat transfer from this insulation system (including conduction, convection and radiation) was 0.86–1.1 mW/(cm K) at an average temperature of 144 K.

Alternative insulation approaches also include: using glass microspheres [22] or Perlite powder in a annular space (either filled with inert gas or vacuum) surrounding the cryogenic vessel. Opacified powder in which metallic flakes are mixed with the insulating powder can provide higher performance due to its ability to reflect back more of the thermal radiation. Recent advances in aerogel [23, 24] have led to its use in thermal blankets for cryogenic systems.

A nice comparison for scaling purposes of various insulation systems is given in Table 1.2 [25]. The heat flux in all cases includes conduction, convection and radiation. Notice that between 300 and 77 K the use of Perlite powder in a vacuum is more effective than just a high vacuum space. This results from the Perlite intercepting some of the thermal radiation heat leak. References [21, 25] provide detailed descriptions of many of these alternative insulation system.

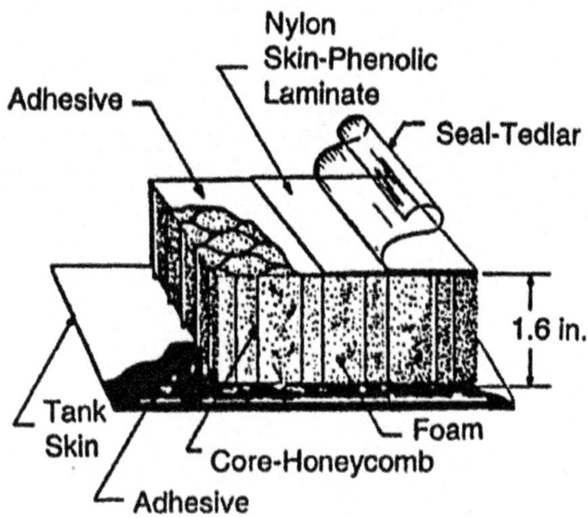


**Fig. 1.10** The foam insulation system of the Saturn 5 s stage $LH_2$ tank [21]

Table 1.2 Comparison of thermal insulation approaches (6 in. thick insulation in all cases)

| Type of insulation | Total heat flux (W/m$^2$) | |
|---|---|---|
| | 300–77 K | 77–20 K |
| Polystyrene foam | 48.3 | 5.6 |
| Gas filled Perlite powder (5–6 lb/ft$^3$ filled with He) | 184.3 | 21.8 |
| Perlite powder in vacuum (5–6 lb/ft$^3$) | 1.6 | 0.07 |
| High vacuum ($10^{-6}$ torr, $\varepsilon = 0.02$) | 9 | 0.04 |
| Opaciated powder (Cu flakes in Santocel) | 0.3 | – |
| MLI | 0.03 | 0.007 |

Reference [25]

## 1.4 Structural Supports for Cryostats

Structural supports for cryostats serve many purposes. An example is the stand or frame that connects the room temperature outer wall of the cryostat to the outside environment (floor, ceiling etc.). Since these devices connect two room temperature components, they are designed using standard structural engineering techniques. Cryostat stands aren't trivial, they may have to support very large loads or react against strong forces (for example from unbalanced pressure loads or magnetic fields) they may need to allow precision movement or alignment of the cryostats.

More interesting for this book are the supports that connect the cryogenic parts of the cryostat with its room temperature outer wall. There are a number of challenges for these supports.

1. They must support the weight of the components plus any applied loads (e.g. pressure, magnetic field, transport, space launch) while at the same time minimizing the conduction heat leak (see Sect. 1.3.1).
2. They must allow for thermal shrinkage of both the components and the supports during cool down.
3. If required, they must achieve the proper alignment of the cryogenic components upon cool down.
4. The supports and the cryogenic components that are attached must be designed to avoid any unwanted resonant frequencies.

The structural supports in the case studies that follow are designed to meet these challenges.

This section first addresses the issue of alignment and then discusses various design solutions for cryostat structures.

### 1.4.1 Alignment Approaches

Some cryostats have little to no alignment requirements. Examples of these include many small laboratory systems, cold boxes for refrigeration plants, transport and liquid storage Dewars. In these cases, the support system must allow the components to shrink and move without causing damage. As mentioned in Sect. 1.2.1 not just supports but all components such as pipes and wiring must be designed to allow for thermal contraction. Over constraining a component by rigidly fixing it so that it can not move may well lead to failure upon cool down.

Many other cryostats do have tight alignment requirements. In these cases, the position of certain components must be in the correct location once at cryogenic temperatures. Examples of these cryostats include: superconducting magnets, superconducting cavities and other components used in particle accelerator beam lines, telescopes and other optical systems and space based systems. As will be seen in the case studies, the alignment requirements can be quite strict with allowable tolerances as small as a fraction of a millimeter. There are three general approaches to meeting alignment requirements.

Maintain the alignment upon cool down. In this approach, the structure is designed so that the cool down is symmetric. Those components with tight alignment requirements are properly aligned while warm and keep this alignment when cooled down. In some cases the alignment of the components changes but stays within the allowable alignment tolerance. Examples of this approach include the ESS cryomodules [26].

Allow the alignment to change in a predictable way. Here the components are allowed to change position during cool down, but do so in a predictable way that is repeated in each cool down. With these designs, the component is installed so that its out of alignment while at room temperature but moves into alignment when cooled. An example of this is the ILC cryomodule (Chap. 5). In this design, it was calculated that the center of the cavity beam tube would shrink upwards by 1.8 mm upon cool down. The cryomodules were aligned into the accelerator beam line so that the center of the beam tube was 1.8 mm lower than the ideal beam line. Upon cool down, the cavity beam tube was properly aligned with the rest of the system.

Allow realignment once cold. In this approach, components with tight alignment tolerances are realigned into the proper position after they have reached their final cryogenic temperature. This can be accomplished via screws or rods that can adjust the position of the cryogenic components from room temperature or by the use of remotely controlled positioners or motors within the cryostat. However, there are a number of disadvantages with this choice. The screws, rods, motors or positioners may add heat leak, may be expensive and may reduce system reliability. While cryostats have been designed in this way, it's generally best to use approach as the last resort or as a back up to the other two approaches.

## 1.4.2 *Suspension of Components from a Room Temperature Top Flange*

This is a very common choice and is used in many laboratory cryostats as well in liquid storage Dewars and refrigerator cold boxes. An example of this approach is shown in Fig. 1.11. All the components inside the innermost part of the cryostat are suspended from the room temperature top flange and not fixed anywhere else. As

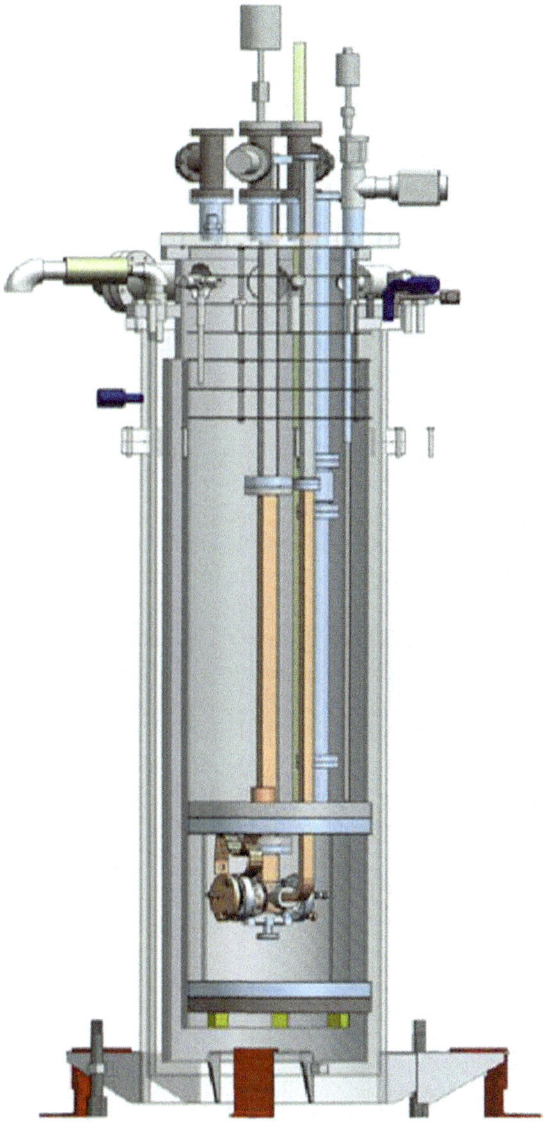

**Fig. 1.11** Cutaway view of a vertical test cryostat at the SLAC National Acceleratory Laboratory

the system is cooled down, the components are allowed to shrink upwards towards the top flange. This design choice does not tend to guarantee the alignment of components and functions best where alignment tolerances are very loose or nonexistent. The suspension of the components also means that changes in the cryostat's orientation or motion during transport or operation may cause the components to oscillate and become damaged. The use of stops to limit the motion or shipping restraints may be required. This approach has the virtues of being simple, low cost and may result in low conduction heat leaks to the cryogenic components. Figure 1.12 shows this design applied to two different refrigerator cold boxes.

When applied to liquid storage Dewars, this design choice results in everything, including the inner vessel and thermal shields, being suspended from the neck of the Dewar. Again, stops or shipping restraints may be required.

This approach can also be taken with accelerator beam line components if the alignment requirements aren't too stringent. Figure 1.13 shows the inner components of the ATLAS Upgrade cryomodule built at Argonne National Laboratory [27]. The Superconducting cavities, magnets, shields and cryogenic lines are all fixed to the upper flange. The movement of the components up as they cool down is within the alignment tolerances.

### *1.4.3 Space Frames*

In this approach, the cryogenic components are attached to a, typically cylindrical, frame that is thermally and structurally attached to the inside of the room temperature vacuum vessel of the cryostat. The frame itself is kept at or near to 300 K and the cryogenic components are connected by long low conductivity metal rods made from stainless steel, titanium etc.

Figures 1.14 and 1.15 show views of the design of the European Spallation Source (ESS) Elliptical Cavity Cryomodule [26]. Note the use of the hanging rods that attach the cavity string, thermal shields and cryogenic pipes to the space frame. The components can all be assembled onto the space frame and then the entire complete assembly can be inserted into the vacuum vessel, aligned properly and then locked into place. The long cylindrical axis of such systems results in all the components cooling symmetrically towards the center with the result that the cavities maintain their original room temperature alignment at cryogenic temperatures. Space frames are very popular in Superconducting Radiofrequency cryomodules and have also been used in the Spallation Neutron Source and the Jefferson Lab 12 GeV Upgrade projects among others (see Chap. 6).

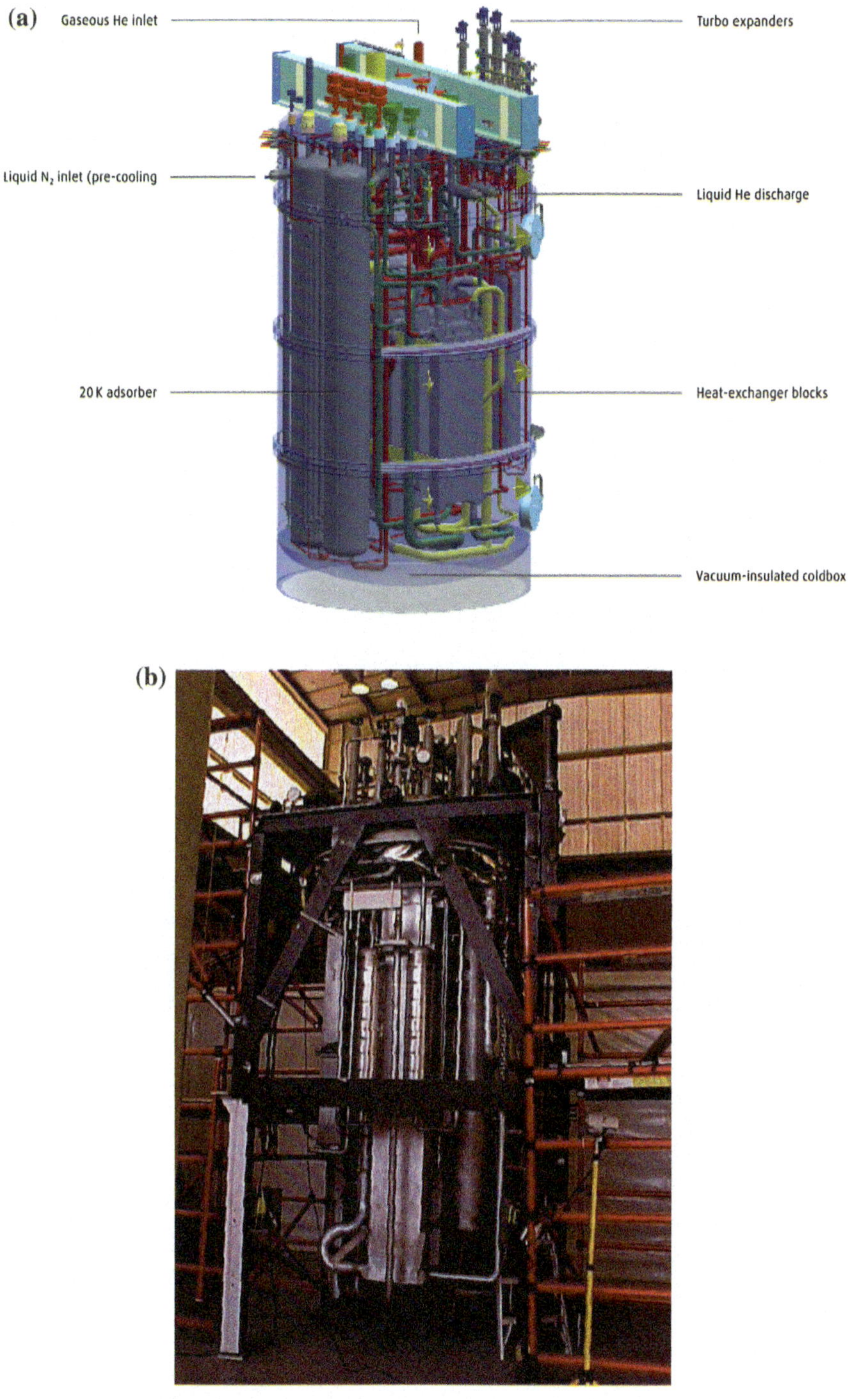
(a)
Gaseous He inlet
Turbo expanders
Liquid N$_2$ inlet (pre-cooling
Liquid He discharge
20 K adsorber
Heat-exchanger blocks
Vacuum-insulated coldbox
(b)

◀ **Fig. 1.12** Views of two cryogenic refrigerator cold boxes showing components suspended from a top plate. **a** Courtesy by Linde Kryotechnik AG and **b** the CTI-4000 cold box at the SLAC National Accelerator Laboratory

**Fig. 1.13** Components suspended from the upper plate of the ALTAS upgrade cryomodule [27]

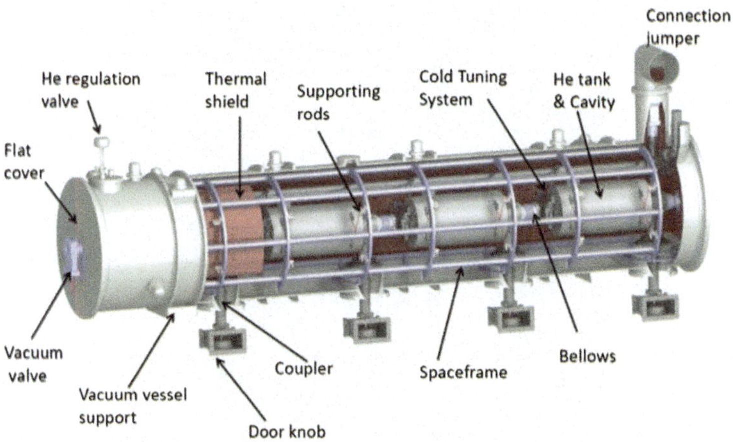


**Fig. 1.14** Cut away view of the ESS elliptical cavity cryomodule. Courtesy of Gilles Olivier (CNRS-IPN Orsay)

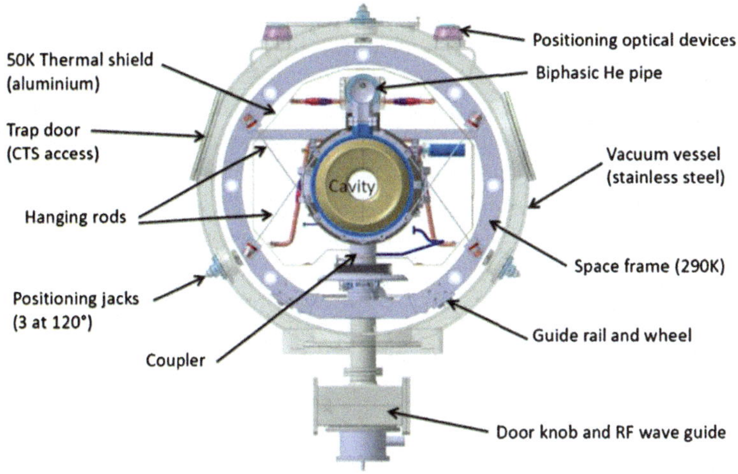


**Fig. 1.15** Cross sectional view of the ESS elliptical cavity cryomodule. Courtesy of Gilles Olivier (CNRS-IPN Orsay)

## *1.4.4 Support Posts*

Cryogenic components can also be connected to the 300 K vacuum vessel via support posts. The components can either sit on top of the support posts or be suspended from the support posts. The posts themselves are designed to minimize the heat transfer between 300 K and cryogenic temperatures (see Sect. 1.3.1). An advantage of this approach is that the posts can be designed to support very large loads and are thus frequently used in magnet systems. Since the cryogenic components are not supported in a symmetric manner the position of the components will change during cool down and this effect must be allowed for in the design. Examples of cryostats using support posts include: the LHC magnet cryostat (Chap. 3), the ILC cryostat (Chap. 5) and the SSC magnet cryostat (Chap. 2).

## *1.4.5 Supports in Space Cryogenics*

Space cryogenic systems have additional challenges in cryostat supports. Not only must they minimize heat lead leak via the supports but the large forces that develop during the rocket launch require (at least temporarily) strong supports. Creative approaches have been developed to solve this problem.

Since the orientation of the launch loads is well known, one approach is to optimize the design by making the supports which resist the launch loads stronger while minimizing the strength of the supports in other orientations. This approach was taken with the AXAF cryostat [15].

Another technique is the use of Passive Orbital Disconnect Struts (PODS). These supports are designed in such a way that during orbit or while stationary on Earth, the cryogenic components are supported on thin wall, low heat leak tubes. However, during launch, the forces are taken up by much stronger, thick wall tubes that are parallel to the thin wall tubes. Once the spacecraft is in orbit, the thick wall tubes passively disconnect and the lower heat leak thin wall tubes again provide the support. One advantage of this approach is that the use of the thin wall tubes while on Earth reduces the heat load during any long launch delays and permits verification of the expected heat load in space. These devices have gone through a number of generations and have been extensively tested and studied [28]. They were a key component of the Gravity Probe B cryostat [29].

## 1.5 Instrumentation

The correct measurement of properties such as temperature, pressure, flow, level and vacuum in cryostats is a key factor in the success of the cryostats. Measurements allow us to understand if our cryogenic components are working properly, enable us to control them and permit the collection of scientific data. There are many subtleties in the selection and installation of cryogenic instrumentation. Poor sensor selection and installation can result in wildly inaccurate readings or sensor failure.

A frequent mistake is to fix one's attention only on the sensor making the measurement itself. By far, the better approach is to think of each measuring point as a complete system including: the sensor, sensor calibration, wiring, feed through, data acquisition hardware and software. The cost of such a complete system can easily exceed $1000 per measurement point. Additional material on cryogenic instrumentation may be found in [8, 30].

### *1.5.1 Temperature Measurement*

Temperature in cryogenics is determined by measuring a physical quantity such as voltage, resistance or pressure that changes in a known repeatable way with temperature. A wide range of options for temperature measurement exists for cryogenics. The existence of reliable commercially available temperature sensors for use in cryogenics has been a major advance in the field. The need to use home made temperature sensors has almost been completely eliminated. The use of home made temperature sensors should be avoided in lieu of commercial devices that are repeatable, reliable, accurate and whose behavior is well understood. The performance of temperature sensors depends on the operating temperature. Some devices, such a silicon diodes, are responsive (though with varying sensitivity) from room temperature all the way down to 1 K. Others, such as platinum resistors, become

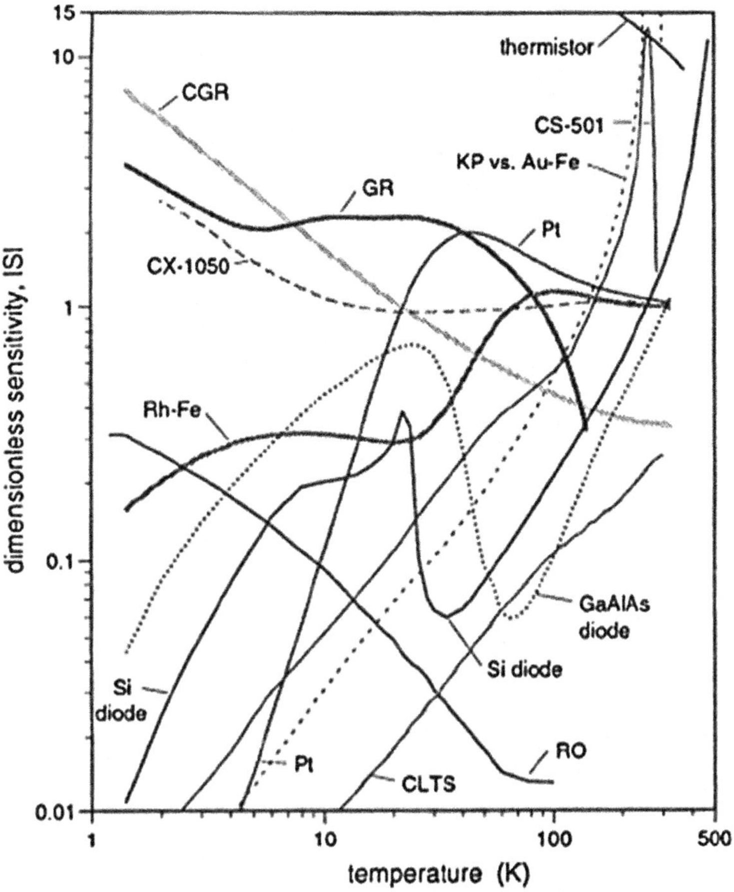


**Fig. 1.16** Dimensionless sensitivity of cryogenic temperature sensors [8]

essentially unresponsive below a given temperature (typically 30 K). Choosing the correct temperature sensor for the temperature range of interest is key. In some cases, a combination of sensors will be required. Figures 1.16 and 1.17 show the sensitivity and temperature resolution of a wide range of temperature sensors.

Self Heating. An issue with temperature sensors, particularly those operating in a vacuum space is that the excitation current applied to the sensor can cause the sensor itself to heat up and thus provide a false measurement. This problem is avoided by always following the manufacturer's recommendations regarding excitation.

Environmental Conditions. The performance of temperature sensors is greatly influenced by the environment in which they operate. Some sensors such as, silicon diodes or germanium resistors perform poorly in magnetic fields or ionizing radiation environments while others such as carbon glass or Cernox sensors work well

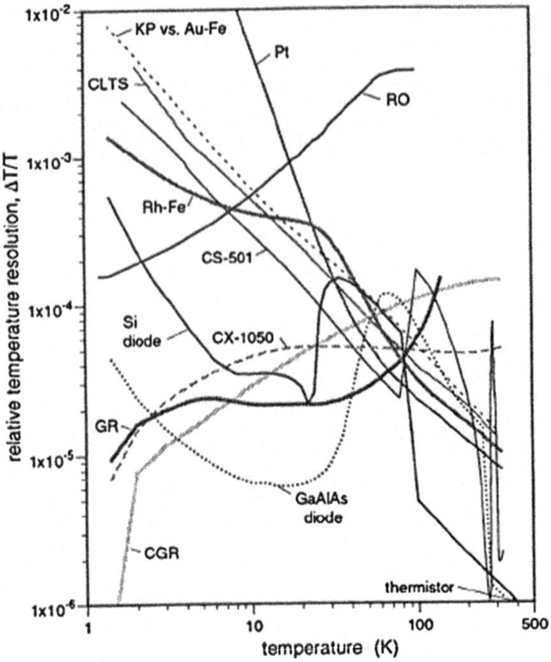


**Fig. 1.17** Relative temperature resolution of cryogenic temperature sensors [8]

under these conditions. Care must be taken to choose the correct sensor for the expected operating conditions. Table 1.3 shows the impact of magnetic and ionizing radiation on common cryogenic temperature sensors. More details may be found in [8].

Calibration. Some types of temperature sensors, such as platinum resistors or silicon diodes have general calibration curves that apply for all sensors of a given type. Others such as cernox or germanium need to be individually calibrated for

**Table 1.3** Impact of magnetic field and ionizing radiation on common cryogenic temperature sensors

| Sensor type | Suitability for use in magnetic fields | Suitability for use in ionizing radiation |
|---|---|---|
| Silicon diode | Poor | Poor |
| Platinum resistor | Poor | Good |
| Germainium resistor | Poor | Poor |
| Carbon glass resistor | Fair | Good |
| Ruthenium oxide | Good | Good |
| Cernox | Good | Good |
| Thermocouple | Good | TBD |
| Cryogenic linear temperature sensors | TBD | Good |

Data from Ref. [8]

best accuracy. In addition, the calibrations of some sensors will drift over time and manufacturers will have recommendations on the need for periodic recalibration of their sensors. In many cases, the calibration drift is not significant, but in situations requiring high accuracy, the cryostat has to be designed to either allow in situ recalibration of the sensors or removal of the sensors for recalibration.

Redundancy. Even the best installed sensors can fail. Once installed inside of cryostats, temperature sensors can be very difficult to remove. In situations where proper temperature measurement is vital to the success of the cryostat, either redundant sensors should be installed at each measuring point or the sensors should be installed in such a way that they can be replaced. This can be accomplished by placing the sensors on removable probes or by installing access ports near the sensors to allow their replacement in the case of in accessible cryostats such as space cryostats, redundant sensors are the answer.

### *1.5.2 Pressure Measurement*

Pressure measurements in cryogenic systems are generally carried out by room temperature pressure sensors. There are a variety of high precision room temperature pressure sensors commercially available. Room temperature pressure transducers have the advantage of being easily replaceable. They are connected to the measuring point via small capillary tubes. These tubes have some disadvantages. First, if improperly designed, they can lead to thermoacoustic oscillations (see Sect. 1.9) and experiments [31] have shown that the capillaries can damp out and delay high speed pressure signals. In those situations where either space requirements or the need for accurate high speed pressure signals exist, there are some pressure transducers that will operate at cryogenic temperatures. Another option is to use non-cryogenic pressure transducers but locate them near the point of measurement in an insulated and heated enclosure to maintain them at room temperature. An example of this approach is given in [32]. Many pressure transducers, both warm and cryogenic, are susceptible to damage from ionizing radiation. Such devices may need local shielding when operated in a radiation environment.

### *1.5.3 Flow Measurement*

As in the case of pressure measurement, flow measurement in cryostat systems is best carried out at 300 K. This increases the amount of flow meters available and allows easier repair or replacement. In cases where flow measurement at cryogenic temperatures is required, many typical flow metering approaches e.g. Venturi meters, orifice plate meters, turbine flow meters etc. can be applied at cryogenic temperatures. Care must be taken to ensure that the meters are properly calibrated for the properties such as density, viscosity, pressure, temperature of the cryogenic fluids. Additional installation requirements such as a certain amount of straight pipe upstream and downstream of the meter must also be met. Many cryogenic flow meters function by measuring a pressure difference and thus the comments on pressure transducers also apply here. Additional information on cryogenic flow metering can be found in [8] with some specific examples given in [33, 34].

### 1.5.4 Level Measurement

Measuring the level of a cryogenic liquid bath is important for proper control and operation of cryogenic systems. There are a number of ways to accomplish this measurement. In liquid helium systems, the use of superconducting level gauges is common. In these devices a probe containing a superconductor is the placed into the bath. The superconducting probe is designed so that the wire is superconducting when in contact with liquid helium and normal conducting when in contact with the corresponding helium vapor. The total voltage drop across the wire can then related to the liquid level. Work [35, 36] has been done to extend this approach to warmer fluids such as hydrogen, nitrogen and oxygen using high temperature superconductors; but alternative methods such as capacitive or pressure differential techniques are more common for these fluids.

In the capacitive approach [37, 38], the differing dielectric constants between the liquid and vapor phases of the fluid are used to determine the fluid level. These are commercial devices and work with a number of cryogenic fluids. As they do not require superconductors to operate they are also useful for fluids warmer than liquid helium.

The differential pressure approach essentially measures the weight of the liquid level of the fluid from which the level can be calculated. Again, these don't require superconductivity.

A final technique, that provides discrete rather than continuous liquid level measurements is the use of Liquid Vapor Detectors (LVDs). These devices can either be discrete superconductors that function similarly to superconducting liquid level probes or more commonly temperature sensors that are powered so that they self heat and provide a very different response in the vapor phase than the liquid phase. Reference [39] provides an example of such devices. This technique was tested in the SHOOT Project (see Chap. 4).

### 1.5.5 Installation, Wiring, Heat Sinking and Feedthroughs

Cryogenic instrumentation systems can easily give inaccurate results if improperly installed. There are a number of issues to take into account.

Keep in mind that wires will generally contract significantly when cooled (see Sect. 1.2.1). If the instrumentation wires are installed with no slack or loops

**Table 1.4** Required heat sinking lengths for wire to achieve a measurement within 1 mK of the actual value

| | | | Heat sinking length (mm) | | | |
|---|---|---|---|---|---|---|
| Material | $T_h$[K] | $T_l$[K] | 0.21 mm$^2$ (24 AWG) | 0.032 mm$^2$ (32 AWG) | 0.031 mm$^2$ (36 AWG) | 0.005 mm$^2$ (40 AWG) |
| Copper | 300 | 80 | 160 | 57 | 33 | 19 |
| | 300 | 4 | 688 | 233 | 138 | 80 |
| Phosphor-Bronze | 300 | 80 | 32 | 11 | 6 | 4 |
| | 300 | 4 | 38 | 13 | 7 | 4 |
| Maganin | 300 | 80 | 21 | 4 | 4 | 2 |
| | 300 | 4 | 20 | 7 | 4 | 2 |
| 304 SS | 300 | 80 | 17 | 6 | 3 | 2 |
| | 300 | 4 | 14 | 5 | 3 | 2 |

Results assume that the wire and sensors are in vacuum and that the thermal conductivity of the adhesive is given by the fit to the thermal conductivity of GE 7031 varnish [8]

between fixed points; upon cool down the wires will shrink and may break or come loose from a solder connection. Always allow sufficient slack in wire installation to allow for this shrinkage.

Instrumentation wires provide a thermal conduction path and thus heat leak between room and cryogenic temperatures. Thus, the number and diameter of the wires along with the type of material in the wires should be chosen to reduce this heat leak. Experience with an early prototype of the ILC Cryomodule (see Chap. 5) illustrates what can happen if this conduction path is not properly considered. Heat sinking the wires at intermediate temperatures (again allowing for shrinkage) will also reduce the heat leak.

An important aspect of heat sinking comes in when using cryogenic temperature sensors in a vacuum space. This is a common installation in which temperature sensors are attached to the outside of a pipe or vessel containing cryogenic fluid. The sensor itself while attached to the cold surface sits inside a vacuum space. Heat leak down the wires, even from an intermediate temperature heat sink can cause the temperature sensor to read an erroneously high temperature despite itself being in good thermal contact with the pipe or vessel. The solution is to always heat sink the wire to the measurement point immediately adjacent to the sensor. Table 1.4 shows the length of wire required to be heat sunk adjacent to the sensor to result in a temperature measurement of within 1 mK of the actual value. As can be seen, the impact of wire choice can be quite impressive. If one were to use 36 gauge Phosphor-Bronze wire, a common choice in cryostat systems, only 7 mm of wire would need to be heat sunk near a sensor operating at a 4 K surface. However, if one were to use 24 gauge copper wire in the same scenario, more than half a meter of wire would need to be heat sunk near the sensor to get a proper reading.

A final installation comment involves the use of feed throughs. If a sensor is immersed in a cryogenic fluid there are basically two options for connecting the sensor via wires to the outside environment. The most common and by far preferable option is to feed the wire through a tube that links the cryogenic fluid environment to the room temperature environment. The room temperature side of the tube is topped with a room temperature electrical feed through that passes the signals to the outside world while maintaining a barrier between the cryogenic space and the outside environment. There are some disadvantages to this approach. The required tube can be hard to route between the two spaces and provides a parallel thermal path from 300 K to the cryogenic space. Such a tube may also be susceptible to thermoacoustic oscillations (see Sect. 1.9).

The second approach, which can be tempting, is to use a cryogenic feed through that connects the cryogenic space to the vacuum space and then run just wires in the vacuum space up to a room temperature connector. This eliminates the tube and along with it, the possibility for thermoacoustic oscillations and the parallel conduction path. It also allows greater flexibility in routing the wires out of the cryostat. Despite its apparent advantages, the use of cold electrical feedthroughs should be avoided if at all possible. The reason is that these cold feedthroughs tend to have poor reliability and can leak. If a leak occurs then there is a leak directly between the cryogenic fluid and the cryostat vacuum space. Such a leak may well spoil the vacuum and stop operation of the cryostat until fixed. Such a repair can be quite time consuming. It's better to use a warm feed through approach. In some cases, such as the SHOOT Helium Dewars (Chap. 4), the use of cold feed throughs is necessary but in theses case extensive testing and prototyping is required to ensure high reliability.

### *1.5.6 Commercial Availability of Instrumentation Systems*

During the past 25 years, the commercial availability of sensors and related instrumentation systems suitable for cryostats and cryogenic systems has significantly increased. This is particularly true for temperature sensors but also applies for other measuring systems. Whenever possible, the best practice is to use commercial solutions. This allows the cryostat designer to benefit from the extensive research, development and testing that have gone into these devices and to take advantage of the customer support provided by these companies. A good source for finding cryogenic instrumentation is the Cryogenic Society of America's annual Buyers Guide [40].

Some instrumentation problems can not be solved by existing devices and research into cryogenic instrumentation continues. Good resources for recent research in cryogenic instrumentation include Advances in Cryogenic Engineering [41], *Proceedings of the International Cryogenic Engineering Conference* [42] and the journal *Cryogenics* [43].

### 1.5.7 Best Practices for Cryostat Instrumentation

Summing up this section, when considering instrumentation in cryostat design, one should keep in mind:

1. Treat instrumentation as a complete system: sensor, wiring, feed through, calibration and data acquisition hardware and software.
2. Don't use more accuracy and precision than required.
3. Use commercially produced sensors whenever possible—there is a lot available.
4. When possible, mount sensors outside cryostat at 300 K (e.g. pressure transducers, flow meters).
5. Play close attention to the installation of wiring and to the proper heat sinking of both wiring and sensors.
6. Consider the possibility of thermoacoustic oscillations. See Sect. 1.9 of this chapter.
7. For critical devices inside of cryostats, install redundant sensors whenever feasible.
8. Be sure to consider how to recalibrate sensors.
9. If at all possible avoid, cold instrumentation feedthroughs.
10. Once prototyping is complete, minimize number of sensors in series production of cryostats.

## 1.6 Seals and Connections

A common feature of cryostats is the need to connect together components such as pipes, vessels, flanges and feedthroughs. In many cases, these connection have to provide vacuum tight seals at cryogenic temperatures; separating a cryogenic fluid from a vacuum space.

Whether the connection will operate at room temperature or cryogenic temperature, the most reliable approach is to weld the components together. Good quality welds, particularly in stainless steel, are the simplest way to create reliable, vacuum leak tight connections. A common temptation is to include many demountable joints, such as flanged connections, in the cryostat design to allow easy assembly and disassembly. This temptation should be avoided. While reliable, demountable, connections do exist; the difficulty in finding and fixing leaks in cryogenic systems is so large that welding and then later cutting systems apart is the better approach. This is particularly true in cryostats that are produced in large numbers such those for the LHC and SSC magnets (Chaps. 2 and 3) or those for the ILC cryomodules (Chap. 5).

Other metal joining techniques include brazing (sometime used in copper systems) and soldering. Soldering of tubes and pipes should be avoided, except

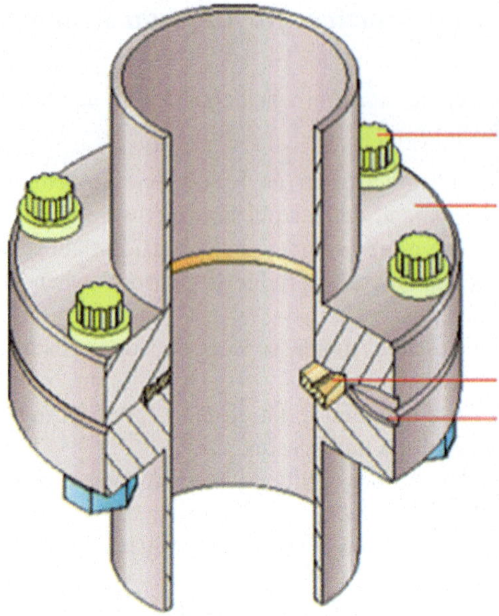


**Fig. 1.18** Example of a ConFlat flange using a soft copper gasket—Courtesy MDC Vacuum Products, LLC

possibly in the case of small laboratory experiments where reliability is less important than the ability to make quick changes. Even in this case, flanged connections are superior if space permits.

When welding dissimilar metals such as aluminum, stainless steel and niobium, transition joints are needed. Reliable transition joints based on brazing, explosive welding or diffusion bonding of the dissimilar metals together have been developed for use at cryogenic temperatures [44, 45] and are available commercially [46].

There are situations where demountable cryogenic connections are required. Two common approaches for this problem are flanged assemblies and bayonets.

Flanged assemblies require a gasket or O-ring to create the final seal between the mating surfaces. At cryogenic temperatures, polymer or rubber O-rings will become brittle and won't work. However, soft metal gaskets (such as copper) will work at cryogenic temperatures. Figure 1.18 illustrates such a system. These devices are reliable and can be obtained commercially [47]. Another flanged option is the use of indium wire for the O-ring. This approach is illustrated in Fig. 1.19. Note the relative areas of the indium wire and Vee grove indicated in the figure. Indium O-ring flanges are generally made in house as opposed to purchased commercially. Flanged connection can be used in all orientations and can separate vacuum spaces, fluids from vacuum and fluids from each other. Care must be taken in flanged connections that differential thermal contraction upon cooling does not result in the bolts holding the flanges together shrinking less than the flanges themselves. This may result in a lessening of the force holding the flanges together resulting in a leaky joint. One solution to this problem is to use Invar washers in the bolt assembly. Invar (see Table 1.1) contracts very little between 300 and 4 K. Properly designed into a

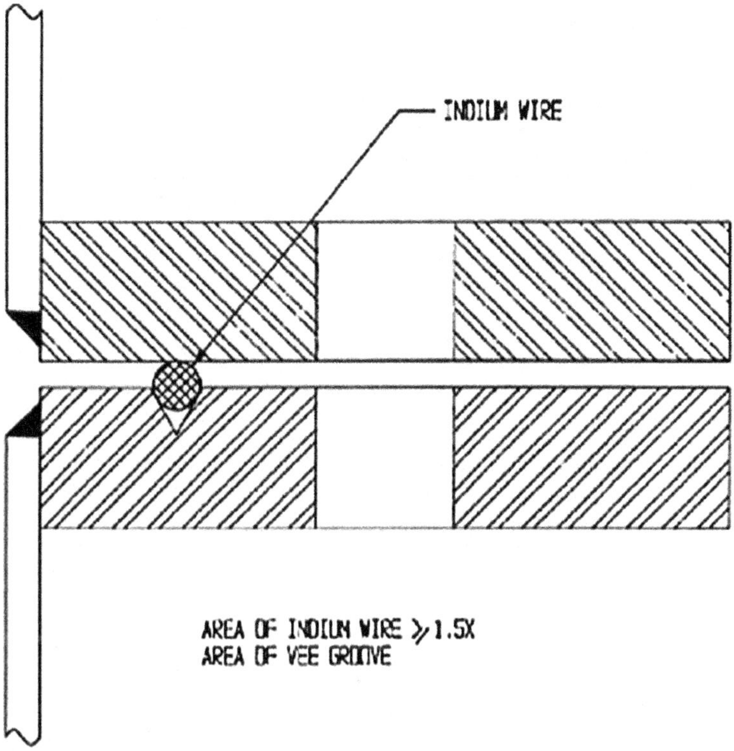


**Fig. 1.19** View of an Indium O-Ring flange [48]

flanged assembly, these washers can serve to keep or increase the force on the flange joint during cool down, preventing leaks. Figure 1.20 and Eq. 1.5, both from Ref. [48] show how to design such a joint assembly using Invar washers.

$$X = [L_A(\alpha_A - \alpha_S) + \varepsilon(L_A + L_S)]/(\alpha_S - \varepsilon - \alpha_I) \quad (1.5)$$

where

$X$ required thickness of the Invar washer
$\alpha_A$ temperature expansion coefficient for aluminum
$\alpha_S$ temperature expansion coefficient for stainless steel
$\alpha_I$ temperature expansion coefficient for Invar
$L_A$ thickness of aluminum flange
$L_S$ thickness of stainless steel flange
$\varepsilon$ unit strain of bolts due to cool down (m/m).

Another common approach to making demountable cryogenic piping connections is to employ bayonets. Bayonets are a set of nested pipes (one "male" and one "female") connected together by a room temperature seal. Figure 1.21 illustrates a typical bayonet pair. Note that the bayonets employ some of the thermal insulation techniques described above; for example, long thin walls connect the room

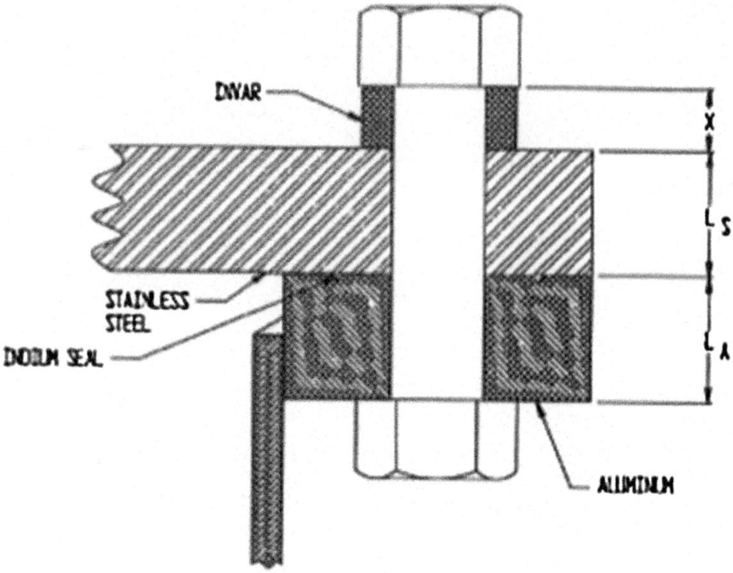


**Fig. 1.20** Flange assembly using Ivar washers [48]

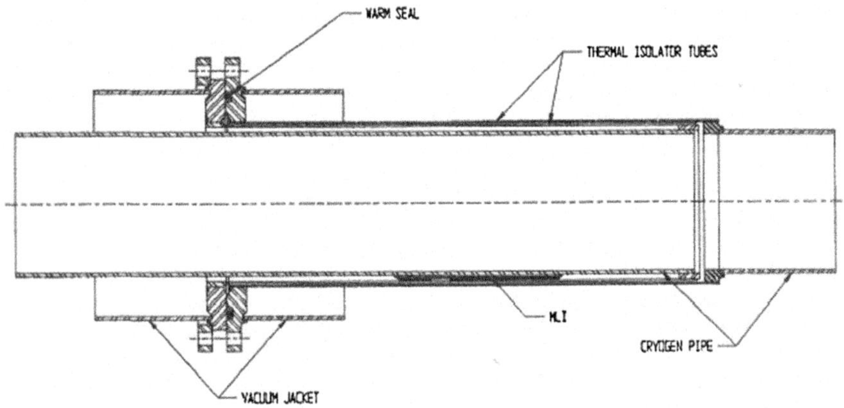


**Fig. 1.21** Example of a cryogenic bayonet pair [48]

temperature and cryogenic portions of the bayonets. The small gap between the female and male bayonets is designed to impede the development of convective heat transfer cells between the cryogenic and room temperature portions thus further reducing heat leak.

Bayonets have a number of advantages, they are commercially available, reliable and can be easily disconnected to decouple different cryogenic components. There are disadvantages as well to bayonets; they are expensive, add additional heat leak

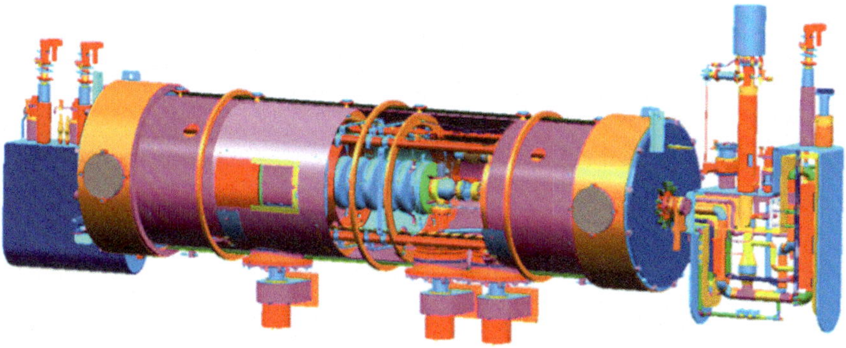

**Fig. 1.22** Cutaway view of the SNS medium beta cryomodule—Courtesy E. Daly-Jefferson Lab, see Chap. 6). Connection to the cryogenic system are done via bayonets on the end boxes

to the system, are not as reliable as welded assemblies and can be hard to disconnect if the pipe diameter becomes too large. Additionally, unlike welds and flanged assemblies, bayonets can only be used in vertical or near vertical orientations. That is, the warm end of the bayonet must be above the cold end. If this is not done, then there is the possibility that the cryogenic fluid will flow into the annual space between the bayonets causing significant heat leak and cooling and possible failure of the room temperature seal.

Bayonets are a valuable tool in cryostat design but are best used only in situations where the regular disconnection of cryogenic components is expected. Bayonet connections are common in segmented cryomodule designs (Chap. 6) where they allow rapid disconnection of individual cryomodules. Figure 1.22 shows an example of bayonets used in a cryomodule design.

## 1.7 Transfer Lines

In large scale systems, cryostats are typically connected to each other, to cryogenic refrigeration plants or to other equipment via insulated transfer lines. These transfer lines move cryogenic fluids between the various components. Transfer lines can be quite complicated and their proper design is an important aspect of a successful cryogenic system. Complicated transfer line systems are also referred to as cryogenic distribution systems. In effect, transfer lines are simply another type of cryostat. Chapter 9 presents more details and examples of cryogenic transfer lines.

Figure 1.23 [49] shows the cross section of a typical multiple line cryogenic transfer line. Notice that it follows the basic principles of thermal insulation. The internal cryogenic pipes are enclosed within an outer vacuum jacket and surrounded by both a actively cooled radiation and an MLI blanket. A low conductivity support separates the pipes and shield and connects them to the outer vacuum jacket.

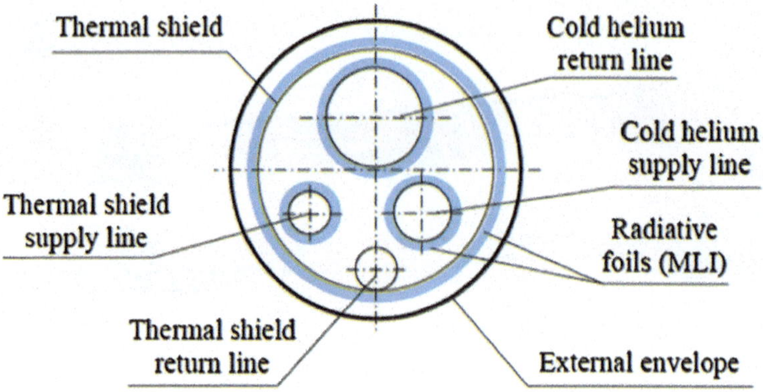


**Fig. 1.23** Cross section of a typical multiline cryogenic transfer line—Courtesy J. Fydrych, ESS (see Chap. 9)

Transfer lines have particular requirements based on their typical geometry. Due to their length, there is significant thermal contraction of the lines during cool down. This contraction has to be dealt with on the design. The common approach is to have some of the internal supports fixed to both the internal pipes and to the outer vacuum jacket with the remainder of the internal supports allowing relative movement between the pipes and the outer vacuum jacket. Bellows are then installed on the pipes between the fixed supports to allow relative contraction of the pipes. An example of this approach is shown in Fig. 1.24 [50]. Note the fixed support and the support that can move relative to the outer vacuum jacket. An alternative approach is to fix all the supports to the vacuum jacket but design some of them so that the pipes can move freely through them upon cool down.

There are other possibilities rather than the use of bellows in the design of transfer lines to allow for thermal contraction. If the transfer line contains enough bends or elbows of sufficiently large bend radius then the motion of the pipes with in the bends relative to the vacuum jacket may be enough to compensate for thermal contraction. Another option is to construct the cryogenic pipes from Invar. As shown in Sect. 1.2.1, this material contracts very little upon cool down. Constructing pipes from Invar greatly reduces the amount of thermal contract that needs to be allowed for in the design, potentially simplifying the design. This approach is described in Refs. [51, 52].

A final issue that must be allowed for in transfer lines is the effect of the line pressure on closed ends of the line. Given the size of many cryogenic pipes, this pressure can result in a very large unbalanced force that must be transferred to the vacuum jacket and from the jacket via support structures to the building or earth. Improper allowance of this effect can result in damage, motion or loss of alignment of the transfer line.

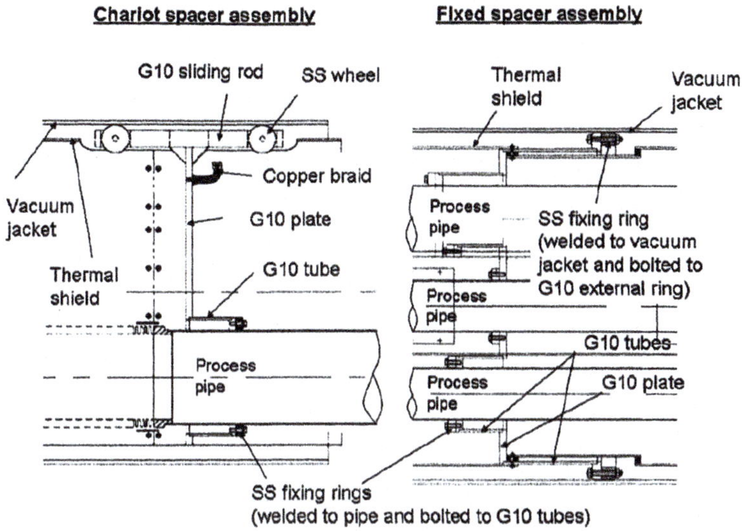


**Fig. 1.24** Example of fixed and moving spacers in a cryogenic transfer design. Reproduced with permission from Parente et al. [50]. Copyright 2006, AIP Publishing LLC

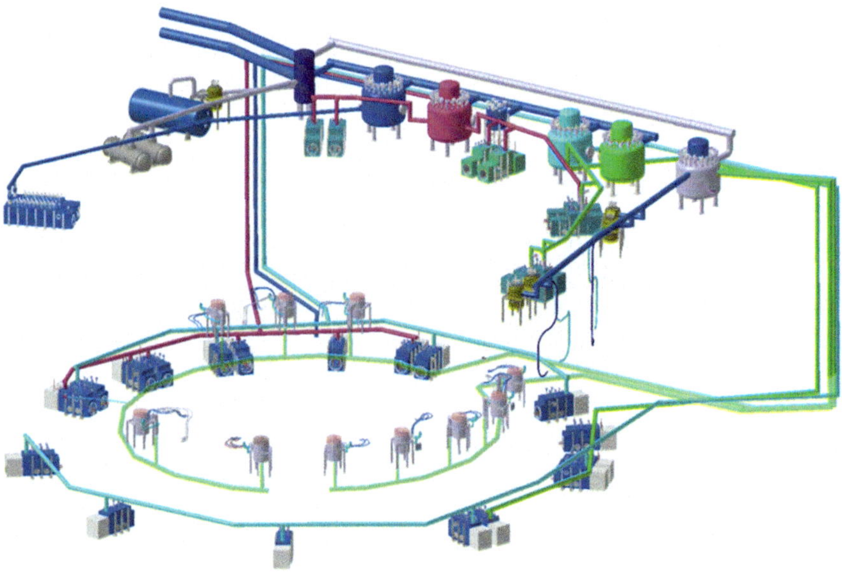

**Fig. 1.25** The ITER cryodistribution system. Reproduced with permission from Serio [53]. Copyright 2010, AIP Publishing LLC

Figure 1.25 [53] is an example of a complex transfer line or distribution system. While most transfer line are custom made, there exist commercial options for simpler systems [54].

## 1.8 Safety

Safety is a broad topic throughout the field of cryogenics. References [55–60] provide a good overview of cryogenic safety. There are safety issues specific to cryostats and these issues should be planned for at the very earliest stage of the design. Altering the design later to remove safety issues can be very time consuming and expensive.

One of the most common mistakes in cryostat design that lead to safety hazards is the use of inappropriate materials. Materials (see Sect. 1.2) that are inappropriate for cryogenic use can become brittle and suddenly fail. The solution here is to always use materials proven for cryogenic service or conduct tests of the material to show that it is suitable for cryogenic temperatures. One should always keep in mind that materials nominally expected to operate at room temperature, may though some other failure, become cooled to cryogenic temperatures and then themselves fail. Such scenarios should be considered in the design of the cryostat.

A hazard that always has to be addressed in cryostat design is over pressurization. This hazard stems from the very large volume difference between a cryogenic liquid at its normal boiling point, i.e. at 1 Bar and the gas phase at room temperature and pressure. Table 1.5 shows this volume ratio for number of cryogenic fluids.

As can be seen, the ratios are very high. Thus, if a cryogenic liquid is in a closed volume and converted to a warmer gas, very high pressures will result. These pressures can easily result in explosive failure of the closed volume causing death, injury and property damage.

The design solution to this problem is to always install properly sized relief valves in any cryostat so that pressures never go above those for which the cryostat is designed.

The details of properly sizing relief valves depend on local requirements, including the applicability of pressure vessel codes. In most locations, it is a requirement that pressure vessels be designed, built and certified in accordance with these codes such as the AMSE Pressure Vessel Code [61] or the European Pressure Directive [62]. Since pressure vessels are frequently defined as vessels in which the maximum allowable working pressure is greater than 1.5 Bar (absolute) then many cryostat vessels fall under these codes. In some cases, such as cryostats for space

**Table 1.5** Volume ratios between cryogenic fluids at their normal boiling point and at 300 K and 1 Bar

| Fluid | $V_{gas}/V_{Liquid}$ |
|---|---|
| Helium | 701 |
| Parahydrogen | 788 |
| Neon | 1341 |
| Nitrogen | 646 |
| Argon | 779 |
| $CO_2$ | 762 |
| Oxygen | 797 |

applications, weight restrictions prevent full implementations of pressure vessel codes. In this case local authorities generally required demonstration of an equivalent level of safety.

While details on pressure vessel requirements and relief valve sizing vary between localities, general guidelines for pressure safety in cryostats include:

1. Always use relief valves that are certified under the applicable pressure code.
2. Never place a shut off valve between the relief valve and the space it is protecting.
3. Consider how any recalibration requirements of the relief valves will be met.
4. Perform a risk analysis of the system. Ask "what if" questions and then install relief valves to cope with resulting hazards. In particular, look for volumes that may not be pressure relieved under certain conditions. Keep in mind that process valves may be operated incorrectly or may leak; that cryogenic systems may warm up suddenly and that vacuum insulation systems may fail.
5. While isolation vacuum jackets typically don't see pressure, they will if the cryogenic system they contain leak. Allow for this eventuality by always installing appropriately sized pressure relief devices on vacuum jackets.
6. Never disable or remove pressure relief devices and never introduce cryogenic fluids into systems without suitable pressure reliefs.
7. Have the cryostat's pressure relief system reviewed by in-house or external experts.

When considering possible failure modes for sizing relief valves, a frequently used worse case scenario is the sudden loss of insulation vacuum. It key to remember here that the issue here is not the sudden appearance of convective heat transfer (Sect. 1.3.2) but rather that the in rush of air will condense on the cryogenic surfaces depositing large amounts of heat and boiling off the cryogenic fluids contained in the vessels and piping. This effect is much more significant than heat deposited by convective heat transfer. The addition of multilayer insulation wrapped around the surface of the cryogenic vessel or pipes helps reduce this effect. Further information is given in [63]. In the case of accelerators with superconducting magnets or superconducting RF cavities, the same effect is seen in the sudden loss of beam tube vacuum [32].

Additional details of pressure rises in magnet cryostats are provided in Chap. 7 of this book. A useful reference for the sizing of relief valves is found in [64].

Cryostats containing oxygen or flammable liquids like hydrogen and LNG have additional unique hazards; always involve experts in these areas when working with such fluids.

As mentioned at the start of this section, safety in cryogenics extends beyond issues associated solely with cryostats. Significant hazards including Oxygen Deficiency Hazards need to be considered. References [55–60] should be consulted for the broader issue of cryogenic safety.

## 1.9 Thermoacoustic Oscillations

Thermoacoustic oscillations (TAOs) are spontaneously occurring pressure oscillations that are frequently seen in cryogenic systems. They generally occur in geometries in which a tube connects room temperature with a cryogenic fluid and is closed at the 300 K end. The tube thus has a strong temperature gradient between the warm and cold ends. Unfortunately, this geometry describes many common cryostat design solutions including: tubes connecting instrumentation wires to room temperature feed throughs, a capillary tubes connecting a cryogenic bath to a room temperature pressure transducer, pressure relief lines and a closed bayonet connections.

"TAOs begin when the temperature gradient causes cold gas in the tube to warm and expand, thus increasing in pressure. This increased pressure then pushes the gas into the colder end of the tube, causing the pressure in the warmer end to fall. The gas then moves back to the warmer end to occupy this now lower pressure space. Under the proper conditions of tube size and temperature gradient, sustained pressure oscillations can be set up" [65]. Thermoacoustic oscillations can result is large pressure swings which can damage equipment and also are very efficient at moving heat between the warm and cold ends of the tube greatly increasing the heat load to the cryogenic fluid.

Ideally, one would like to design systems to avoid TAOs. While this is not completely possible, there have been studies on idealized systems that provide some guidance. The physics behind TAOS indicate that there should be two zones of stability where TAOS won't occur. If the tube is small enough, then the friction of the oscillating gas on the tube wall will damp out the oscillations preventing the TAOS from starting. This is known as viscous damping. The other area of stability occurs in large tubes or vessels where the mass of the fluid is too big for the fluid to be moved significantly by the heat being transferred form the warm tube wall and thus oscillations will not start. This is know as inertial damping.

Experimental studies by Gu [66] on the idealized case of tubes with linear temperature gradients have shown exactly these regions of stability. Figure 1.26 shows these results. For each set of parameters, the left hand region of stability represents the viscous damping while the right hand region of stability represents the viscous damping. While idealized these curves can at least give some guidance. The parameters on the plots are defined as:

$$\alpha = \frac{T_H}{T_C} \tag{1.6}$$

$$\xi = \frac{L_H}{L_C} \tag{1.7}$$

where

$T_H$ the warm temperature

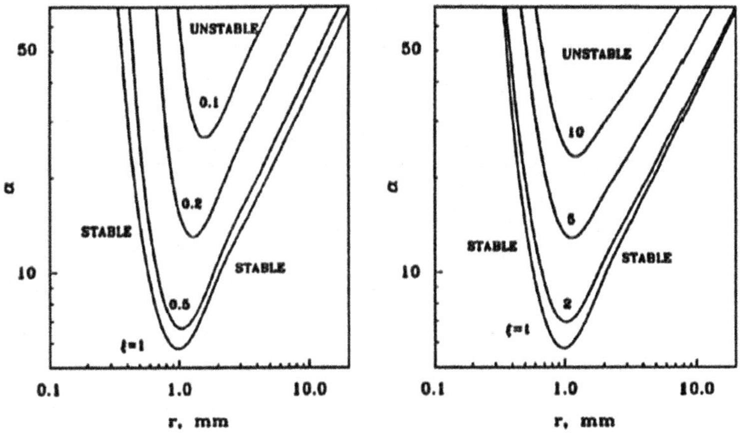


**Fig. 1.26** Stability maps for thermoacoustic oscillations [67]

$T_C$ the cold end temperature
$L_H$ the warm section length
$L_C$ the cold section length

The limit between the cold and warm lengths of the tube is defined as the point in the linear temperature gradient where

$$T = (T_H + T_C)/2 \tag{1.8}$$

The studies were done with a 1 m long tube. In order to apply the results to other geometries, use the reduced radius r′ defined as:

$$r' = r/\sqrt{L} \tag{1.9}$$

The studies described above used a very ideal case and thus even if you are in the stable regions you may find TAOs occurring. There are mitigations that can be taken in the event that TAOs are present. These include:

- Add volume to the warm end or design the warm end with fixtures to allow the addition of this volume
- Install a check valve between the warm and cold end (near boundary between the two)—this converts the problem to a closed cold tube with no TAOs
- Heat sink the closed end (thus changing $T_H/T_C$)
- Allow flow through the warm end.

Further details on TAOs may be found in Refs. [67–70].

## 1.10 Prototyping and Series Testing

A common aspect in the creation of successful cryostats is the building and testing of a prototype. This process allows the performance of the design to be tested and compared against requirements. It is not uncommon for the results of the prototype testing to lead to changes, sometimes major changes, in the final cryostat design. Prototyping is found in almost all of the case studies that follow in this book. Full scale prototypes, tested under the expected final operating conditions provide the most value but in some cases, testing of prototype subassemblies is sufficient.

For systems in which a number of identical cryostats are required, series testing of the final production cryostats is also recommended. Testing the performance of completed cryostats under their final operating conditions provides the best evidence that the cryostats are being properly manufactured. Due to the complexity of many cryostats, small changes in the manufacturing process can lead to significant problems such as vacuum leaks, electrical shorts or higher than designed heat loads. Series testing of the completed cryostats as they are produced can identify these problems in time to correct them in the remaining cryostats under production. Series testing was carried out for the LHC magnet cryostats [71], the SRF cryomodules for XFEL [72] and 12 GeV Upgrade [73]. Such testing is also planned for the LCLS II [74] and ESS [75] cryomodules.

The cost and schedule impact of both prototyping and series testing can be significant; requiring expensive test stands and months of additional time. There is always a temptation to skip these steps and move directly to a test of the final integrated system that uses the cryostats. Doing so, however, leads to a significant risk that the original design may not meet specifications or that manufacturing errors have resulted in poor performance. If this occurs, the cost and schedule impact of repairing the cryostats can easily be more than the initial impact of any forgone prototyping and series testing. Generally speaking, it is advisable to allow for prototyping and series testing in the initial cryostat project plan.

## References


1. V. Ganni, P. Knudsen, Helium refrigeration considerations for cryomodule design. Adv. Cryo. Eng. **59B** (2014)
2. J.G. Weisend II, C. Pagani, R. Bandelmann, D. Barni, A. Bosotti, G. Gygiel, R. Lange, P. Pierini, D. Sellman, S. Wolff, The TESLA test facility (TTF) cryomodule: a summary of work to date. Adv. Cryo. Eng. **45A** (2000)
3. F. Haug, T. Peterson, J.G. Weisend II, *Cryogenic Safety—A Guide to Best Practice in the Lab and Workplace* (Springer, USA, in Preparation)
4. J.G. Weisend II, R. Boyce, W. Burgess, A. Candia, R. Carr, J. Gao, K. Gustafsson, C. Jones, W. Kaminskas, G. Oxoby, R. McKeown, H. Quack, A. Scott, T. Weber, The cryogenic system for the SLAC E158 experiment. Adv. Cryo. Eng. **47A** (2002)
5. R.P. Reed, A.F. Clark (eds.), *Materials at Low Temperatures* (American Society of Metals, 1983)

6. J.G. Weisend II, V. Flynn, E. Thompson, R.P. Reed, A reference guide for cryogenic properties of materials. SLAC-TN-03–023, web information available at http://www.slac.stanford.edu/cgi-wrap/getdoc/slac-tn-03-023.pdf. Accessed 7 Jan 2015
7. T.F. Durham, R.M. McClintock, R.P. Reed, *Cryogenic Materials Data Handbook* (US Dept. of Commerce, National Bureau of Standards, Washington, DC, 1961)
8. J.G. Weisend II (ed.), *The Handbook of Cryogenic Engineering* (Taylor & Francis, New York, 1998)
9. G. Ventura, M. Perfetti, *Thermal Properties of Solids at Room and Cryogenic Temperatures* (Springer, New York, 2014)
10. Cryocomp, Eckels Engineering. http://www.eckelsengineering.com/
11. METALPAK, CPAK and EXPAK from Cryodata. http://www.htess.com/software.htm
12. H.M. Rosenburg, *Low Temperature Solid State Physics* (Oxford University Press, Oxford, 1963)
13. T.H. Nicol, TESLA test cell cryostat support post thermal and structural analysis. Fermilab-TM-1794 (1992)
14. S.W. Van Sciver, *Helium Cryogenics* (Springer, New York, 2012)
15. S.R. Breon, R.A. Hopkins, S.J. Nieczkoski, The X-ray spectrometer—a cryogenic instrument on the advanced X-ray astrophysics facility. Adv. Cryo. Eng. **37B** (1992)
16. A. Poncet, V. Parma, Series-produced helium cryostats for the LHC magnets: technical choices, industrialisation, costs. Adv. Cryo. Eng. **53A** (2008)
17. W.L. Johnson, A.O. Kelley, J.E. Fesmire, Thermal degradation of multilayer insulation due to the presence of penetrations. Adv. Cryo. Eng. **59A** (2014)
18. J.G. Weisend II, Cryogenic engineering, in *Mechanical Engineer's Handbook*, ed. by M. Kutz (Wiley, New York, 2015)
19. J. Fesmire, S. Augustynomicz, B.E. Scholtens, Robust multilayer insulation for cryogenic systems. Adv. Cryo. Eng. **53B** (2008)
20. J. Polinski, M. Chorowski, A. Choudhury, T. Datta, Synthesis of the multilayer cryogenic insulation modeling and measurements. Adv. Cryo. Eng. **53B** (2008)
21. T.M. Flynn, *Cryogenic Engineering* (Marcel Dekker, New York, 1997)
22. R.G. Baumgartner, E.A. Myers, J.E. Fesmire, D.L. Morris, E.R. Sokalski, Demonstration of microsphere insulation in cryogenic vessels. Adv. Cryo. Eng. **51B** (2006)
23. S. White, J. Demko, A. Tomich, Flexible aerogel as a superior thermal insulation for high temperature superconductors. Adv. Cryo. Eng. **55A** (2010)
24. J.E. Fesmire, S. White, G. Gould, S. Augustynowicz, Aerogel blanket insulation materials for cryogenic applications. Adv. Cryo. Eng. **55B** (2010)
25. R. Barron, *Cryogenic Systems* (McGraw-Hill, 1966)
26. G. Oliver, J.P. Thermeau, P. Bosland, G. Devanz, F. Lesigneur, C. Darve, ESS Cryomodule for elliptical cavities, in *Proceedings of the 16th International Conference on RF Superconductivity* (2013)
27. J.D. Fuerst, S.M. Gerbick, M.P. Kelly, M. Kedzie, et al. Assembly, installation and commissioning of the ATLAS upgrade cryomodule. Adv. Cryo. Eng. **55A** (2010)
28. R.T. Parmley, Passive orbital disconnect strut (PODS-III), structural and thermal test program. NASA CR 166473 (March 1983)
29. R.T. Parmley, Unique cryogenic features of the gravity probe B experiment. Adv. Cryo. Eng. **33** (1988)
30. J.W. Ekin, *Experimental Techniques in Low-Temperature Measurements* (Oxford University Press, Oxford, 2006)
31. F. Haug, A. McInturff, Measurement of pressure transmission in long capillaries. Adv. Cryo. Eng. **35** (1990)
32. R.C. Dhuley, S.W. Van Sciver, Sudden vacuum loss in long liquid helium cooled tubes. IEEE Trans. Appl. Supercond. **25**, #3 (2015)
33. S.W. Van Sciver, D.S. Holmes, X. Huang, J.G. Weisend II, He II flowmetering. Cryogenics **31**, 75–86 (1991)

34. T. de Jonge et al., Development of a mass flowmeter based on the Coriolis acceleration for liquid, supercritical and superfluid helium. Adv. Cryo. Eng. **39** (1994)
35. K. Kajikawa et al., Fundamental investigation of a superconducting level sensor for liquid hydrogen with $MgB_2$ wire. J. Phys. Conf. Ser. **97** (2008)
36. R. Kurunanithi el al., Calibration of an HTS based LOX 400 mm level sensor, in *Proceedings of ICEC 25: Physics Procedia* (2015), p. 67
37. I.V. Velichkov, V.M. Drobin, Capacitive level meters for cryogenic fluids with continuous read-out. *Cryogenics* **30** (June 1990)
38. R. Sawada et al., Capacitive level meter for liquid rare gases. Cryogenics **43**, 449–450 (2003)
39. M.J. DiPirro, A.T. Serlemitsos, Discrete liquid/vapor detectors for use in liquid helium. Adv. Cryo. Eng. **35** (1990)
40. 2015 Buyer's Guide (Cryogenic Society of America, 2015). http://www.cryogenicsociety.org/buyers_guide/
41. Advances in Cryogenic Engineering **1–62** (1955–2016)
42. *Proceedings of the 25th ICEC, Physics Procedia* **67** (2015). http://www.sciencedirect.com/science/journal/18753892/67
43. *Cryogenics*. http://www.journals.elsevier.com/cryogenics/
44. J.D. Fuerst et al., Niobium to stainless steel braze transition development, in *Proceedings of the 11th Workshop on RF Superconductivity*
45. W. Soyars et al., Superfluid helium testing of a stainless steel to titanium piping transition joint. Adv. Cryo. Eng. **55A** (2010)
46. An example of commercially available transition joints can be found at: http://www.rbdh.com/Bimetallic-transition-joints,164.html
47. See, for example, ConFlat flanges from various manufacturers
48. G. McIntosh, Cryostat design, in *The Handbook of Cryogenic Engineering*, ed. by J.G. Weisend II (Taylor & Francis, New York, 1998)
49. E. Pyata et al., XFEL injector-1 cryogenic equipment. Phys. Procedia **67**, 868–873 (2015)
50. C. Parente et al., The local helium compound transfer lines for the large hadron collider cryogenic system. Adv. Cryo. Eng. **51** (2006)
51. S. Claudet et al., Two 100 m invar transfer lines at CERN: design principles and operating experience for helium refrigeration, in *Proceedings of ICEC 20* (2005)
52. A. Thakar et al., Design and analysis of a bellows free cryogenic transfer line, in *Proceedings of International Conference on Current Trends in Technology* (2011)
53. L. Serio, Challenges for ITER cryogenics. Adv. Cryo. Eng. (2010)
54. An example of commercially available transfer lines may be found at: http://www.nexans.de/eservice/Germany-en/navigate_315900/Cryogenic_Systems.html
55. F.J. Edeskuty, W.F Stewart, *Safety in the Handling of Cryogenic Fluids* (Springer, 1996)
56. M.G. Zabetakis, *Safety with Cryogenic Fluids* (Plenum Press, 1967)
57. F.J. Edeskuty, M. Daugherty, Safety, in *The Handbook of Cryogenic Engineering*, ed. by J.G. Weisend II (Taylor & Francis, 1998)
58. T.M. Flynn, Safety with cryogenic systems (Chap. 10), in *Cryogenic Engineering* (Marcel Dekker, 1997)
59. *Cryogenic Safety Manual: A Guide to Good Practices* (British Cryogenics Council, 1991)
60. J.G. Weisend II (ed.), *Cryogenic Safety—A Guide to Best Practice in the Lab and Workplace* (Springer, in preparation)
61. Boiler & Pressure Vessel Code 2015 (ASME, 2015)
62. Pressure Equipment Directive, PED 2014/68/EU
63. C. Heidt, S. Grohmann, M. Sußer, Modeling of the pressure increase in liquid helium cryostats after failure of the insulating vacuum Adv. Cryo. Eng. **59B** (2014)
64. EN 13648-1:2008, Cryogenic vessels. Safety devices for protection against excessive pressure. Safety valves for cryogenic service
65. J.G. Weisend II, Thermoacoustic oscillations, in *Cold Facts* Winter 2013 (2013)
66. Y. Gu, Thermal Acoustic Oscillations in Cryogenic Systems, PhD Thesis, University of Colorado, 1993

67. Y. Gu, K.D. Timmerhaus, Experimental verification of stability characteristics for thermal acoustic oscillations in a liquid helium system. Adv. Cryo. Eng. **39** (1994)
68. T. Yazaki et al., Stability limit for thermally driven acoustic oscillation. Cryogenics (July 1979)
69. T.J. Miller, Y. Gu, Elimination of thermal acoustic oscillations in cryogenic pumps. Adv. Cryo. Eng. **51** (2006)
70. N. Dittmar et al., Onset of thermoacoustic oscillations in flexible transfer lines for liquid helium, in *Proceedings of ICEC 25—Physics Procedia* **67** (2015)
71. V. Chohan et al., LHC magnet tests: operational techniques and empowerment for successful completion
72. M. Wiencek et al., Tests of the accelerating cryomodules for the European X-ray free electron laser, in *Proc. of SRF 2013* (2013)
73. J. Hogan et al., 12 GeV upgrade project—cryomodule production, in *Proceedings of IPAC 2012* (2012)
74. LCLS-II Final Design Report (LCLSII-1.1-DR-0251-R0), 22 Nov 2015
75. W. Hees et al., The ESS cryomodule test stand, in *Proceedings of ICEC 25—Physics Procedia*, vol. 67 (2015)

# Chapter 2
# SSC Collider Dipole Cryostat

**Thomas H. Nicol**

**Abstract** The Superconducting SuperCollider (SSC) project required almost 8000 dipole magnet cryostats. These cryostats had design requirements that included low heat leak, structural stability and cost minimization. Detailed designs were developed and numerous prototype magnets were built and tested. While the SSC project did not continue, many aspects of the SSC dipole magnet cryostats were used in later projects such as the LHC and ILC. This chapter describes the SSC dipole magnet cryostat design including design criteria, thermal insulation systems, structural supports and interconnects. Test results from prototypes, particularly in the area of heat leak are also presented.

## 2.1 Introduction

The design of the Superconducting Super Collider (SSC) required the construction of 7680 superconducting dipole or bending magnets, each 17 m long, 1356 quadrupoles or focusing magnets, each 3.3 m long as well as several special magnets, spool pieces, and other hardware. The entire accelerator was to be housed in an 83 km-long underground tunnel [1].

Fermilab was involved in the design and production of cryostats for all prototype SSC dipole magnets built beginning in 1986. Late in 1989 a fundamental change in the dipole coil design; increasing the physical aperture from 40 to 50 mm, resulted in the need to redesign the dipole cryostat. The larger aperture gave rise to an increase in diameter of the overall cold mass assembly of approximately 25 % and a weight increase of approximately 50 % both of which impact the cryostat design significantly.

Cryostat designs for superconducting magnets are largely driven by thermal and structural considerations. Designers must continually be cognizant of the heat load

T.H. Nicol (✉)
SRF Development Department, Technical Division, Fermi National Accelerator Laboratory, P.O. Box 500, Batavia, IL 60510, USA
e-mail: tnicol@fnal.gov



J.G. Weisend II (ed.), *Cryostat Design*, International Cryogenics Monograph Series,
DOI 10.1007/978-3-319-31150-0_2

to the helium system and of the structural loads imposed on the cryostat systems from static weight, shipping and handling, quench loads, and ambient ground motion. These two considerations are generally at odds with one another. Low heat load implies a minimum of structural material conducting heat from the environment. Sound structural design implies material with sufficient strength to resist both static and dynamic forces.

This chapter summarizes the results of the 50 mm collider dipole cryostat design effort. Thermal and structural aspects of the design of each of the major cryostat systems will be described in detail. The 40 mm dipole cryostat purposely served as the starting point for this design work. Wherever possible, development work on that and other similar cryostats was borrowed in order to take advantage of proven technology. References to the 40 mm cryostat will be made for comparison wherever appropriate. Each of the cryostat systems will be addressed in turn; vacuum vessel, 80 and 20 K thermal radiation shields, multilayer insulation (MLI), cryogenic piping, suspension system, and magnet interconnect.

To begin the design process, one needs a good handle on the pertinent allowable heat loads to each thermal station as well as a good definition of the structural environment to which the magnet will be subjected. Given the large number of magnets, even a small heat load, especially to 4.5 K, could result in a significant increase in the overall capacity of the cryogenic plants needed. Table 2.1 summarizes the thermal and structural design criteria for the 50 mm dipole that affect the cryostat design. Figure 2.1 illustrates a cross section through the 50 mm cryostat [2, 3].

**Table 2.1** Thermal and structural design criteria

| | 4.5 K | 20 K | 80 K |
|---|---|---|---|
| *Static heat loads* | | | |
| Infrared (W) | 0.053 | 2.335 | 19.1 |
| Support conduction (W) | 0.160 | 2.400 | 15.8 |
| Interconnect (W) | 0.150 | 0.320 | 2.1 |
| Total static (W) | 0.363 | 5.055 | 37.0 |
| *Dynamic heat loads* | | | |
| Synchrotron radiation (W) | 2.169 | | |
| Splice heating (W) | 0.140 | | |
| Beam microwave (W) | 0.195 | | |
| Beam gas (W) | 0.136 | | |
| Total dynamic (W) | 2.640 | | |
| Total dipole (W) | 3.003 | 5.055 | 37.0 |
| *Structural load summary* | | | |
| Cold mass weight | | 11,360 kg | |
| Shipping and handling | | 2.0 g | Vertical |
| | | 1.5 g | Axial |
| | | 1.0 g | Lateral |

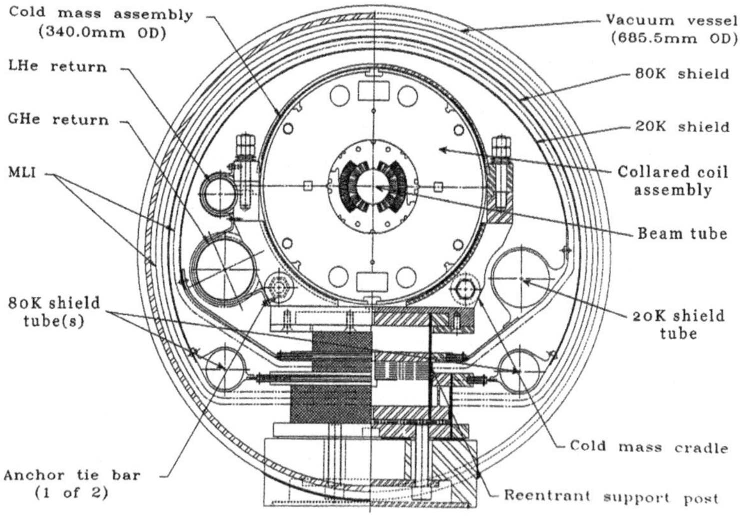


**Fig. 2.1** SSC 50 mm collider dipole cryostat—cross section

## 2.2 Vacuum Vessel

The vacuum vessel was the outermost cryostat component and served to contain the insulating vacuum. In addition, it functioned as the major structural element to which all other systems were ultimately attached to the accelerator tunnel floor. Furthermore, it served as a pressure containment vessel in the event of a failure in an internal cryogen line. The vessel was a carbon steel cylinder 685.5 mm OD, 7.92 mm wall with an overall length of 14,922.5 mm.

The vessel had five penetrations at which the internal cold mass supports were attached. The internal supports were positioned to minimize the deflection of the cold assembly due to its own weight and were spaced on 3179.76 mm centers starting at the center of the magnet. Two external supports served to tie the vessel to the tunnel floor. The external supports were spaced to minimize the deflection of the vacuum vessel due to its own weight and to the weight of the internal components. They were located 4154.17 mm on either side of center.

The 7.92 mm wall of the vessel was not sufficient to transfer suspension system loads to ground immediately around the internal support locations. At these positions a 19.05 mm thick, 559 mm long reinforcement transferred internally generated loads, primarily from shipping and handling, from the magnet support system to the vacuum vessel. Figure 2.2 illustrates the major features of the vacuum vessel and notes applicable dimensions. The vacuum vessel material was low carbon steel. Ordinarily one would like to use a material with greater fracture toughness in the event of a major cryogen spill inside the cryostat, however, cost mandated a more readily available material. In addition, failure analysis indicated that a high fracture toughness material was not required [2, 3].

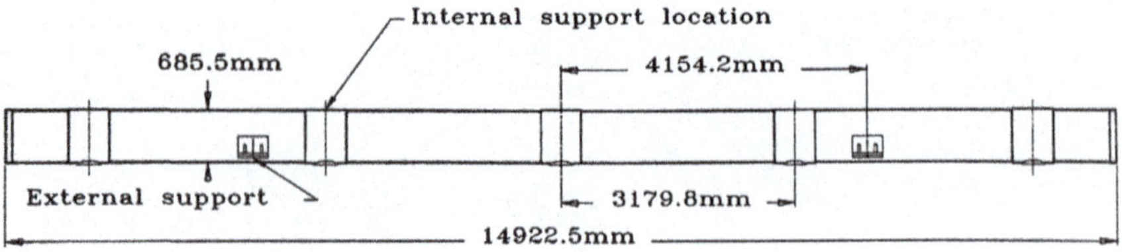


**Fig. 2.2** SSC 50 mm collider dipole vacuum vessel layout

**Fig. 2.3** SSC 50 mm collider dipole complete magnet

Referring only to drawings, it's easy to lose sight of the scale of these magnets. Figure 2.3 shows a complete dipole magnet being transported within the Fermilab site on a full-length tractor-trailer.

## 2.3 Thermal Radiation Shields

In order to minimize thermal radiation heat load to the 4.5 K magnet, the cryostat used two thermal radiation shields. The outer shield was cooled by $LN_2$ and operated nominally at 80 K. The inner shield was cooled by helium gas returning to

**Fig. 2.4** Lower half of the 20 K thermal shield assembly

the refrigerator and operated nominally at 20 K. Each shield was cooled by discrete pipes connected to formed shells which totally encompassed the internal cold assembly. The 20 K shield used a single extrusion attached to the shell. The 80 K shield used two; one serving as the shield supply, the other as the return. This scheme required that the shells have high thermal conductivity to minimize thermal gradients around their circumference. Copper and aluminum are the materials of choice for this application. Thermal analyses on both shields were performed using both materials. The results from these analyses indicated there is no significant difference in the thermal performance of either shield between these two materials. Material cost dictated the ultimate material choice. Copper and aluminum are approximately the same cost per unit weight, however, with a density over three times that of aluminum, the use of copper as the shield material would have resulted in an assembly cost three times higher than an aluminum shield. The selected material for both 80 and 20 K shields was 6061-T6 aluminum, 1.59 mm thick. The shields were segmented along their length to minimize the effect of bowing during cool down caused by the asymmetry of the cooling tubes with respect to the shield shells. The thermal shields are shown in Fig. 2.1. The lower half of a 20 K thermal shield assembly is shown in Fig. 2.4 [2, 3].

## 2.4 Multilayer Insulation

The design requirements for the thermal insulation system in the SSC dipole cryostat dictated that heat load from thermal radiation and residual gas conduction be limited to the values listed in Table 2.1. Essential to meeting these requirements was an insulation system design that addressed transient conditions through high layer density for improved gas conduction shielding, had sufficient mass and heat capacity to reduce the effects of thermal transients, and was comprised of materials suitable for extended use in a high radiation environment. Finally, the system design was such that fabrication and installation techniques guaranteed consistent thermal performance throughout the entire accelerator.

The insulation system needed to have a mean apparent thermal conductivity of $0.76 \times 10^{-6}$ W/cm-K in order to meet the design heat load budget. This was achieved by using a multilayer insulation (MLI) system comprised of reflective layers of aluminized polyester separated by layers of spun-bonded polyester spacer. The reflective layers consisted of flat polyester film aluminized on both sides to a nominal thickness not less than 350 Å. The spacer layers consisted of randomly-oriented spun-bonded polyester fiber mats. The mean apparent thermal conductivity of an MLI blanket comprised of these materials was measured to be $0.52 \times 10^{-6}$ W/cm-K [6, 7]. The MLI system for the SSC 50 mm collider dipole cryostat consisted of full cryostat-length assemblies of MLI fabricated and installed as blankets on the 80 and 20 K shields [2–4].

The MLI system for the 80 K thermal shield consisted of two 32-reflective-layer blanket assemblies, for a total of 64 reflective layers. The stack height of each 32-layer blanket was 8.86 mm, with a mean layer density of 3.61 layers per mm. The blanket design incorporated 32 reflective layers of double-aluminized polyethylene terephthalate (PET) film. The reflective layers were separated by single spacer layers of 0.10 mm spun-bonded PET material. Single layers of 0.23 mm spun-bonded PET covered the blanket top and bottom and served to position the polyester hook and loop fasteners at the blanket edges. The fasteners were affixed to the cover layers by sewing. A third 0.23 mm PET layer was located midway through the blanket assembly and separated the upper and lower 16 reflective layers of MLI. The multiple blanket layers were sewn together as an assembly along both edges of the blanket. Non-lubricated polyester thread was used in all sewing operations.

The MLI system for the 20 K thermal shield consisted of ten reflective layers of PET film aluminized with a nominal coating of 600 Å per side. The reflective layers were separated by three spacer layers of 0.1 mm spun-bonded PET material. Single layers of 0.23 mm spun-bonded PET covered the blanket top and bottom and served to position the polyester hook and loop fasteners at the blanket edges. The fasteners were attached to the cover layers by sewing. The multiple blanket layers were sewn together as an assembly along both edges of the blanket.

At each blanket edge, the upper MLI layers were sewn together from the upper cover layer through to the middle three layers with thread terminated in the three

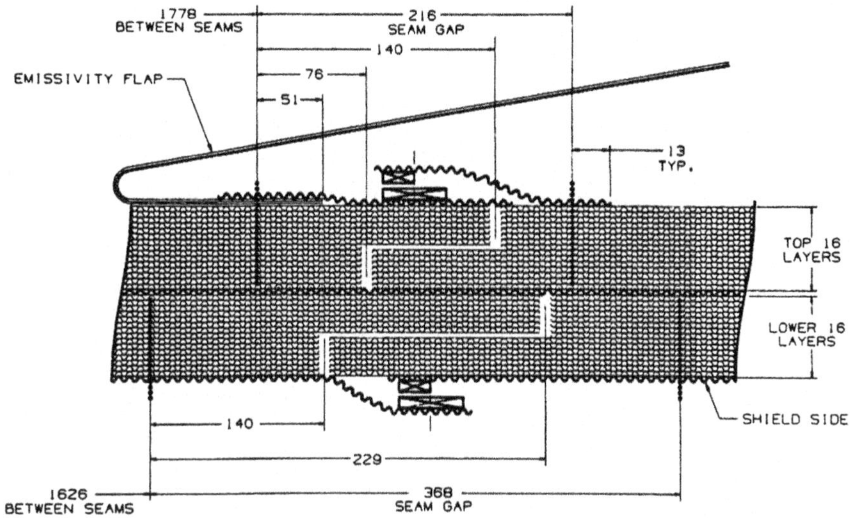


**Fig. 2.5** Cross section through the 80 K MLI blanket

layers. The seam location was then incremented 7.62 cm laterally along the mid-layer and the lower MLI layers were sewn together from the middle three layers through to the lower cover layer. Figure 2.5 shows a cross section of the 80 K MLI blanket.

To facilitate assembly and all the blankets required to support the SSC pre-production program, a system was devised to wrap and sew complete blanket assemblies. It consisted of a wheel with a width equal to the perimeter of the blankets and a circumference equal to their length. The reflector and spacer materials were fed off supply reels onto the wheel and the edges were sewn after wrapping the required number of layers was complete. Figure 2.6 shows the assembly wheel during the winding process.

## 2.5 Cryogenic Piping

In addition to providing the necessary structural support and thermal insulation for the cold mass assembly, the cryostat served to contain the piping for all of the cryogenic services required for magnet and magnet system operation. All the cryogenic pipes are shown in Fig. 2.1. Table 2.2 provides a summary of the pressure and flow parameters for each of these services. Table 2.3 lists the pipe sizes for each of the cryostat pipes. Each pipe was anchored at the center of the magnet assembly, and was free to slide axially at the remaining support points to allow for thermal contraction during cool down.

**Fig. 2.6** MLI blanket winding and sewing apparatus

**Table 2.2** Cryogenic system pressure and flow parameters

| System | P(max) (atm) | Fluid | P(oper) (atm) | v ($cm^3/g$) | T (K) | Flow (g/s) | ΔP (psi) |
|---|---|---|---|---|---|---|---|
| 20 K shield | 18 | He | 2.5 | 163.4 | 20 | 100 | 0.016 |
| 80 K shield | 18 | $LN_2$ | 5.0 | 1.286 | 84 | 750 | 0.012 |
| Single phase | 18 | He | 4.0 | 7.255 | 4.25 | 100 | 0.010 |
| LHe return | 18 | He | 3.0 | 7.609 | 4.425 | 85.6 | 0.010 |
| GHe return | 9 | He | 0.9 | 73.07 | 4.3 | 26.8 | 0.00053 |
| Vac vessel | 2 | He | 0.0 | na | 300 | na | na |

Table 2.3 Cryostat pipe sizes

| System | Pipe ID (mm) | Pipe OD (mm) | Flow area ($mm^2$) | Material |
|---|---|---|---|---|
| 20 K shield | 82.55 | 88.90 | 5352 | Aluminum |
| 80 K shield | 57.15 | 63.50 | 2565 × 2 | Aluminum |
| LHe return | 45.14 | 47.62 | 1600 | Stainless steel |
| GHe return | 86.41 | 88.90 | 5864 | Stainless steel |

## 2.6 Suspension System

The suspension system in any superconducting magnet serves as the structural attachment for all cryostat systems to the vacuum vessel, which in turn anchors them to the accelerator tunnel floor. Conventional suspension systems were effectively reinvented during early SSC magnet development. The intention in development of the 50 mm dipole cryostat was to take advantage of that earlier work tailoring it only for the revised thermal and structural design parameters. Schedule requirements for the redesign work did not allow continued suspension R&D and, in fact, the thermal and structural performance of the 40 mm design did not indicate it to be necessary [2, 3, 5–7].

Using the thermal and structural parameters outlined in Table 2.1 the suspension system for the new design was extrapolated from its 40 mm counterpart. The emphasis was on meeting the allowed suspension system conduction heat load, satisfying the structural requirements, and maximizing the lateral suspension stiffness. This latter constraint is not explicitly defined in the design criteria but arose out of concerns during testing of 40 mm prototypes that the lateral natural frequency was too closely coupled with resonances found in an over the road shipping environment.

As shown in Fig. 2.7, the suspension system consisted of two major components; reentrant style support posts and anchor tie bars. There were five support posts located along the length of an SSC cold mass assembly spaced 3179.76 mm apart centered about the middle of the cold mass. Five being the number that limited the cold mass sag due to self-weight to the allowed 0.25 mm. The support posts resisted vertical and lateral loads imposed during shipping and handling. The cold mass was allowed to slide axially with respect to all but the center post to allow for thermal contraction of the cold assembly during cool down. This implied that, given no other restraint, the center support would need to resist axial loads. A single support was not capable of resisting this potential 1.5 g load. Using a special support at the center that could handle this load would have imposed inordinately high heat loads on the refrigeration system. Rather, a means was developed to distribute axial loads to all five supports without impacting the conduction heat load. Axial tie bars were used to connect the top of each support post to its neighbor(s). In this scheme, an axial load acted first at the center support and then distributed to all five supports.

The tie bars needed to be dimensionally stable when cooled from their assembly temperature of 300 K to their operating temperature of 4.5 K, otherwise their

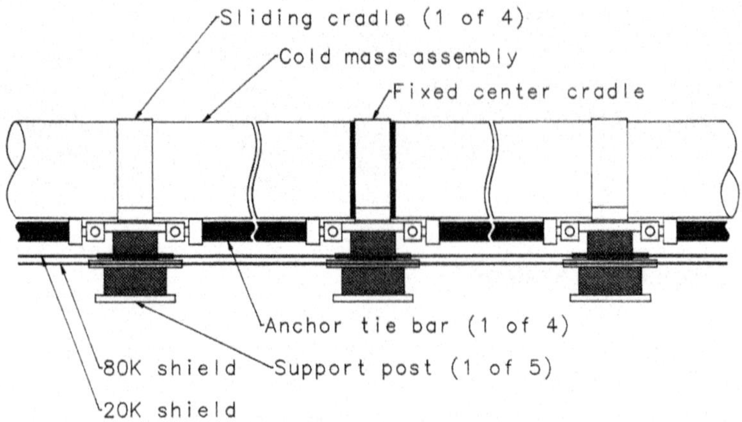


**Fig. 2.7** SSC collider dipole cryostat suspension system components

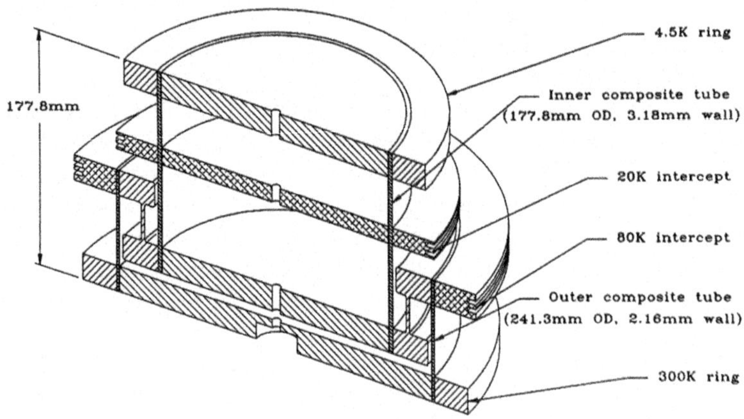


**Fig. 2.8** SSC collider dipole cryostat re-entrant support post

contraction would impose high bending loads on the support posts. Uniaxial graphite fibers in an epoxy matrix provided the solution. This material is extremely stiff and yet exhibits virtually no change in length during cool down. An elastic modulus of 82.5 GPa can be achieved in such a composite even with relatively low modulus fibers.

The configuration of the support posts for the 50 mm cryostat is shown in Fig. 2.8. The outer composite tube was 2.16 mm thick and filament wound using S-glass in an epoxy matrix. The inner tube was 3.18 mm thick and filament wound using graphite fibers in an epoxy matrix. The ability of this design to use two

**Table 2.4** Re-entrant support post design parameters

| | |
|---|---|
| Design lateral load | 32960 N |
| Design vertical load | 22240 N |
| Overall support height | 177.80 mm |
| Outer composite tube | 241.30 mm OD, 2.16 mm wall |
| | 90.55 mm long, glass/epoxy |
| | Design stress: 132 MPa |
| Inner composite tube | 177.80 mm OD, 3.18 mm wall |
| | 150.88 mm long, graphite/epoxy |
| | Design stress: 154 MPa |
| Theoretical lateral natural frequency | 9.41 Hz |
| Heat load to 80 K | 3.133 W |
| Heat load to 20 K | 0.492 W |
| Heat load to 4.5 K | 0.033 W |

different materials allowed each material to be used in its optimum temperature range. Table 2.4 provides a summary of the critical design and performance parameters of the 50 mm cryostat supports.

Figure 2.9 shows the cold mass and a cradle and support post assembly.

## 2.7 Interconnect

As its name implies the magnet interconnect served as the connection area between magnets at which each of the cold mass and cryostat pipes between magnets were connected. Each pipe was anchored axially at the center of the magnet which means they contracted between 25 and 32 mm depending on whether they were stainless steel or aluminum. The relative contraction at the interconnect was then 50–64 mm, i.e. twice the single magnet value. Bellows were required on each cold mass, shield, and cryogenic pipe to allow for this expansion. All bellows were hydro-formed stainless steel. For the aluminum extrusions on the 80 and 20 K shield pipes an aluminum to stainless steel transition joint was required for the bellows connection. Joints using diffusion bonding or brazing between these two materials had been successfully employed in 40 mm prototypes. Lateral instability was a concern for interconnect bellows and required the use of internal squirm protectors on each bellows assembly. During individual magnet testing, the bellows assemblies employed metal seals and mechanical connections, but for tunnel installation, all bellows and pipe connections were welded [2, 3].

Just like for the magnet as a whole, radiative heat transfer needed to be minimized in the interconnect area. This was accomplished by shield bridges that spanned the 80 and 20 K shield gap between adjacent magnets. These bridges were essentially extensions to the magnet shields, modified to contain their respective

**Fig. 2.9** Cold mass and cradle and support post assembly

bellow OD's if necessary. A sliding joint between bridge sections on adjacent magnets accommodated contraction during cool down. Each was covered with the same MLI scheme used throughout the body of the magnet. These shield bridges also contained pressure reliefs for each shield to prevent pressure buildup within either shield in the event of an internal piping failure. The reliefs were located in the upper half of the shield sections in order to prevent liquid spills from impinging directly onto the vacuum vessel wall. Figure 2.10 illustrates a typical magnet interconnection. The external vacuum bellows and thermal shield bridges have been removed for clarity.

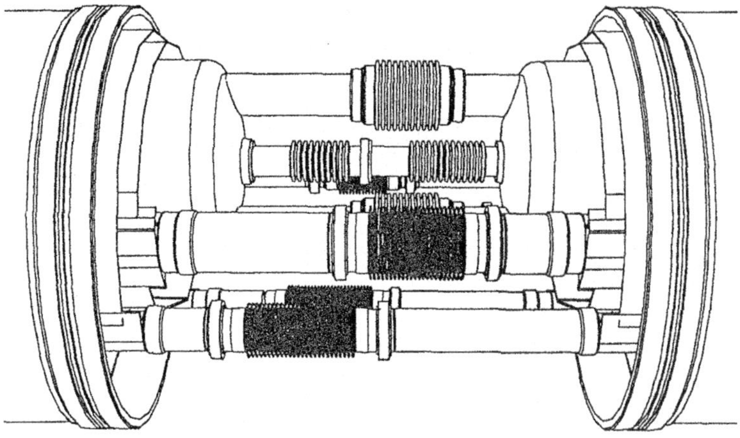

**Fig. 2.10** SSC 50 mm collider dipole magnet interconnect

## 2.8 Test Results

Early in the 50 mm magnet development program a full scale thermal model was constructed and tested in a series of tests at Fermilab. The objectives of the thermal model program were to evaluate the cryostat design from a production standpoint, to gain experience in magnet handling and transportation, to monitor the transient thermal and structural responses of the cryostat, and to measure heat leaks to the cold mass and thermal shields. The thermal model was identical to later magnetic models except that the cold mass assembly contained a simulated collared coil assembly and the model ends were reconfigured for open cycle heat leak measurements. The model was instrumented to evaluate thermal and structural performance and included temperature sensors to monitor cool down, warm-up, and steady state conditions and strain gauges to monitor the performance of the suspension system during onsite transit, cool down, and operation. Construction of the thermal model utilized the same fabrication facility, assembly procedures, and components as the magnetic models [8, 9].

The measurements were performed in an open cycle mode with supply reservoirs at each end of the cold mass. Figure 2.11 shows an overview of the measurement system. The inner shield was supplied from an external dewar. The inner 20 K shield temperature was an operational variable which was controlled by varying its flow rate. The outer 80 K shield was supplied by $LN_2$ reservoirs at each end of the measurement system. The external piping was equipped to measure flow rates of the cold mass and inner and outer shield liquid and gas streams.

After cool down, initial filling, and stabilization of the three systems, the temperature of the inner shield was regulated to a selected test value. After regulation, temperatures and heat leaks were monitored to establish equilibrium, often taking several days. Once at equilibrium, steady state heat leak data was taken.

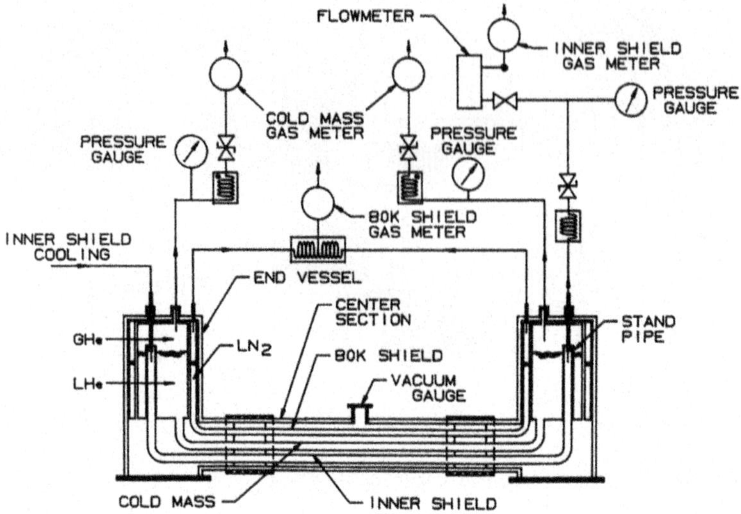


**Fig. 2.11** Thermal model measurement schematic

The cold mass heat leak was determined by boil-off measurements. Since the measurement included the heat leak to the cold mass and the reservoirs at each end of the magnet, a separate measurement called the zero-magnet test was conducted using just the helium reservoirs, which was then subtracted from the total system heat leak. The inner shield heat leak was computed from the shield flow rate and the temperature rise in the shield gas stream. Like the cold mass, the heat leak to the 80 K shield was determined using the liquid boil-off rate. As with the cold mass, the zero-magnet test results were subtracted from the total.

The measurement program took place over approximately a three month period and involved 23 data-taking runs as identified by the inner shield temperature. The following results have been screened to exclude transient periods, upset conditions, and operational problems.

Cool down was gradual due to the open cycle nature of the operation. The cold mass was initially cooled and filled with liquid nitrogen, evacuated, and then filled with liquid helium. The total time required to cool and fill the cold mass with liquid helium was 294 h.

The support post bending loads due to differential axial thermal contraction were low with the exception of the downstream end post which indicated a load of 4800 N. A probable cause for such a load is a non-operational cold mass slide, most likely the result of binding in the slide material itself. The support post temperature profiles agreed well with those of an identical post measured in a heat leak test facility.

Random mechanical noises occurred during cool down and throughout the measurement program likely the result of relative motion between the cold mass,

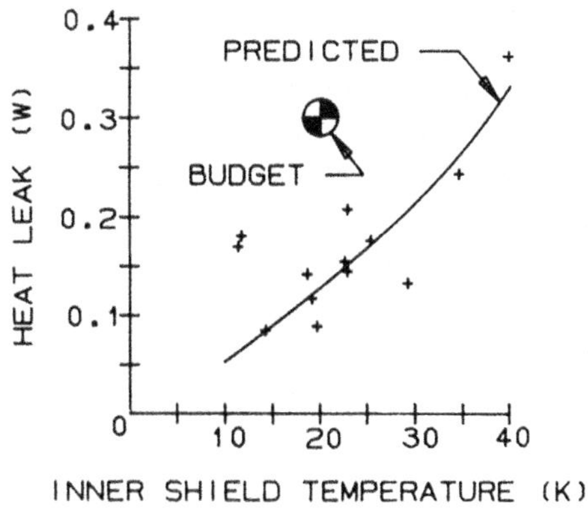


**Fig. 2.12** Cold mass heat leak versus inner shield temperature

thermal shields, and support posts and the vacuum vessel due to ambient temperature variations and settling of the test pad.

The subtractive heat leak contribution of the end reservoirs was measured with the inner shield cooling tube filled with liquid helium to eliminate conduction and thermal radiation to the cold mass end to include end shine thermal radiation. Under these conditions, the shield temperature was 7.6 K and the measured background was 865 mW. The background was not strongly dependent on reservoir liquid level.

Calibration heaters in the cold mass were employed to evaluate the accuracy of the measurement system. The heaters were energized with the inner shield at approximately 20 K. The increase in the measured cold mass heat leak corresponding to heater power levels of 200 and 398 mW were 197 and 376 mW, respectively.

The cold mass heat leak at the cryostat design point with the inner shield at 20 K, was 140 $\pm$ 40 mW as compared to the predicted 128 mW and budgeted 300 mW. The cold mass heat leak versus inner shield temperature is shown in Fig. 2.12.

Considerable differences exist between measured and predicted heat leaks at several experimental points. Factors that can contribute to these differences are as follow:

- The center section heat leak (140 mW) is small relative to the balance of the apparatus (865 mW). This unbalance amplifies end vessel effects.
- Thermal communication exists between the inner shield cooling circuit and the end vessel helium reservoirs as a result of the inner shield supply and return piping passing through the cold mass helium reservoirs. Even though the piping was insulated, transients in inner shield supply conditions could be seen to affect the reservoirs.

Fig. 2.13 Inner shield heat leak versus inner shield temperature

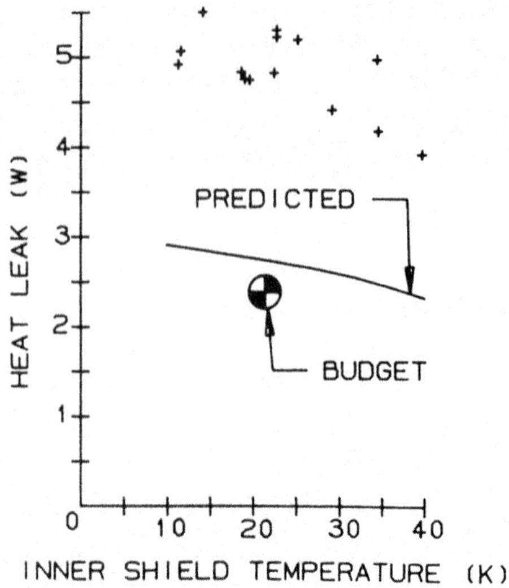


- Level instability (sloshing) occurred occasionally in the end vessels. The sloshing was most often associated with changes in the inner shield circuit operation.
- The predicted sensitivity to insulating vacuum is high. At the design inner shield operating temperature of 20 K, the residual gas (helium) conduction at $10^{-6}$ torr was 43 mW and at $10^{-5}$ torr was 430 mW. The insulating vacuum at the vacuum vessel mid-span port ranged from $1.3 \times 10^{-6}$ to $2.9 \times 10^{-6}$ torr during the span of the measurements. The location and installation of the vacuum gauge was demonstrated by diagnostic measurements to inaccurately relate changes in the insulating vacuum that could correspond to changes in the outgassing rate of the mild steel vacuum vessel due to changes in ambient temperature. Consequently, vacuum was not monitored frequently during most of the data taking.
- Variations in atmospheric pressure result in temperature changes of the liquid which influence the apparent heat leak. Atmospheric pressure was not monitored frequently during most of the data taking. Where possible, the heat leak was corrected for changes of liquid temperature and pressure with time. The cold mass end to end temperature variation during operation was characteristically about 30 mK.

The inner shield heat leak at the cryostat design point was $5.0 \pm 0.4$ W compared to the predicted 2.77 W and budgeted 2.5 W. The inner shield heat leak versus inner shield temperature is given by Fig. 2.13.

A corroboration of the inner shield heat leak measurement was provided by performing inner shield boil-off measurements during the cold mass background measurement. With the shield cooling tube filled with liquid helium, the heat leak as measured by boil-off was 5.84 W.

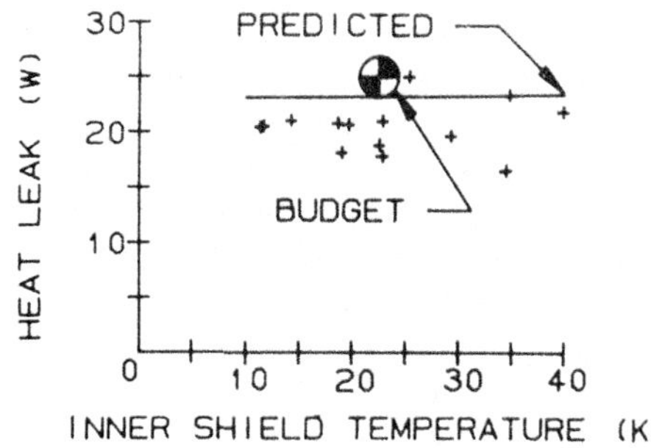


**Fig. 2.14** Outer shield heat leak versus inner shield temperature

The factor of two difference between measured and predicted heat leaks was felt to be due to thermal shorts between the 20 and 80 K systems, by locally compressed insulation between the shields and by an insufficient number of layers of insulation. Shield and support temperature monitors indicate the possibility of shorts. An autopsy of the thermal model to investigate the existence of such thermal shorts was not conducted for programmatic reasons.

The subtractive heat leak contribution of the end reservoirs was made with the center section removed and the reservoirs connected together. The nominal background heat leak was 24 W. The background was found to vary with level. The outer shield heat leak at the cryostat design point was 19 $\pm$ 2 W compared to the predicted 23.3 W and the budgeted 25 W. The outer shield heat leak versus inner shield temperature data is given by Fig. 2.14. The less than predicted measured heat leaks were felt to be associated with the thermal shorts that are suspected to exist between the inner and outer shields and their connections. The scatter in the measured heat leak was felt to be associated with the sensitivity of the insulation systems to insulating vacuum in the higher pressure ranges, i.e. greater than $10^{-4}$ torr. As noted earlier, the insulating vacuum, while not accurately measured, showed changes with ambient temperature.

Following the completion of several prototype 40 and 50 mm aperture dipole magnets, similar, more extensive tests were conducted on strings of magnets at the SSC Laboratory in Waxahachie, TX. Similar discrepancies between measured, predicted, and budgeted heat leaks were documented in those tests as well, but the project was discontinued before more exhaustive tests and analyses could be conducted.

Figure 2.15 shows a completed SSC collider dipole magnet on the test stand in the Fermilab Magnet Test Facility.

**Fig. 2.15** Complete SSC dipole assembly on the Fermilab test stand

## 2.9 Summary

The SSC development program afforded us the opportunity to extend the design of cryostats for superconducting magnets far beyond the state of the art present at the end of the Fermilab Tevatron program. Advances in new materials technology opened up options for cryostat designers in both thermal and structural materials. Strict limits on allowable heat load forced us to develop new mechanisms for structural support and thermal shielding. The end result was a cryostat design which met the demands of the SSC and which served as the starting point for the development of other magnet systems far into the future.

It is gratifying to see projects like the LHC (Chap. 3), TESLA, LCLS-II (Chap. 5), and many others around the world, taking advantage of developments in multilayer insulation technology and fabrication techniques, thermal shield design, cryogenic support structures, and interconnect design that were made during the R&D activities, prototyping, and production of SSC collider dipole magnets.

## References


1. SSC Central Design Group (1986), Superconducting super collider conceptual design. SSC-SR-2020, March 1986
2. T.H. Nicol, Design development for the 50 mm superconducting super collider dipole cryostat. *Supercollider 3* (Plenum Press, New York, 1991), pp. 1029–1036
3. T.H. Nicol, SSC 50 mm collider dipole cryostat design. in *Advances in Cryogenic Engineering*, vol. 37A (Plenum Press, New York, 1991), pp. 525–533
4. J.D. Gonczy, W.N. Boroski, R.C. Niemann, J.G. Otavka, M.K. Ruschman, C.J. Schoo, A blanket design, apparatus, and fabrication techniques for the mass production of multilayer insulation blankets for the superconducting super collider. in *Advances in Cryogenic Engineering*, vol. 35A (Plenum Press, New York, 1990), pp. 507–516
5. T.H. Nicol, R.C. Niemann, J.D. Gonczy, SSC magnet cryostat suspension system design. in *Advances in Cryogenic Engineering*, vol. 33 (Plenum Press, New York, 1988), pp. 227–234
6. T.H. Nicol, R.C. Niemann, J.D. Gonczy, A suspension system for superconducting super collider magnets. in *Proceedings of the Eleventh International Cryogenic Engineering Conference* (Butterworth & Co., Surrey, UK, 1986), pp. 533–538
7. E.T. Larson, J.A. Carson, T.H. Nicol, R.C. Niemann, Improved design for a SSC coil assembly suspension connection. in *Advances in Cryogenic Engineering*, vol. 33 (Plenum Press, New York, 1988), pp. 235–241
8. R.C. Niemann, W.N. Boroski, J.D. Gonczy, T.H. Nicol, J.G. Otavka, M.K. Ruschman, SSC dipole magnet cryostat thermal model measurement results. in *Advances in Cryogenic Engineering*, vol. 33 (Plenum Press, New York, 1988), pp. 251–258
9. J.D. Gonczy, W.N. Boroski, T.H. Nicol, R.C. Niemann, J.G. Otavka, M.K. Ruschman, *Advances in Cryogenic Engineering*, vol. 33 (Plenum Press, New York, 1988), pp. 243–250

# Chapter 3
# Twenty-Three Kilometres of Superfluid Helium Cryostats for the Superconducting Magnets of the Large Hadron Collider (LHC)

**Philippe Lebrun**

**Abstract** The Large Hadron Collider (LHC) at CERN is the world's largest scientific instrument. The 1600 "high-field" superconducting magnets that make up the 23 km circumference accelerator ring represent the largest use of superfluid helium (He II) to date. This chapter describes the design evolution of the LHC magnet cryostats with particular emphasis on the He II cooling system, thermal insulation system and structural supports. Prototype testing, series production, installation and commissioning of these cryostats is also discussed. Numerous figures and tables illustrate the cryostat and present performance results.

## 3.1 The LHC and Its Cryogenic System

The Large Hadron Collider (LHC) at CERN, the European Organization for Nuclear Research near Geneva, is the largest research instrument ever built, and the most advanced tool in elementary particle physics (Fig. 3.1): its two combined synchrotrons, 26.7 km in circumference, accelerate and bring into collision intense counter-rotating beams of protons and ions at high energy, to probe the structure of matter and study the forces of nature at the unprecedented scale of tera-electron-volt (TeV) per elementary constituent [1]. This energy, about thousand times the mass of the proton, corresponds to the temperature conditions in the early universe $10^{-10}$ s after the Big Bang: the LHC therefore constitutes a powerful time machine, in particular to explore the mechanism of electro-weak symmetry breaking which occurred in the very early universe and to search for possible constituents of dark matter, relics from this period. In terms of length scale, 1 TeV is equivalent to $10^{-18}$ m, and the LHC may thus also be seen as an ultra-microscope resolving dimensions thousand times smaller than the proton. Exploration of this *terra incognita* has already led to the discovery of the long-sought Higgs boson in 2012,

P. Lebrun (✉)
CERN, European Organization for Nuclear Research, Geneva, Switzerland
e-mail: Philippe.Lebrun@cern.ch

 
J.G. Weisend II (ed.), *Cryostat Design*, International Cryogenics Monograph Series,
DOI 10.1007/978-3-319-31150-0_3

**Fig. 3.1** Outline of the Large Hadron Collider in the Geneva area (CERN photo)

and is continuing with precision measurement of its properties as well as with search for new physics, "beyond the Standard Model".

To guide and focus its rigid beams along the accelerator tunnel, the LHC uses more than 1600 high-field superconducting magnets operating in 80 t of superfluid helium at 1.9 K (Fig. 3.2). In addition to enhancing the performance of the Nb-Ti superconductor by lower temperature operation, the LHC magnet cooling scheme exploits the unique thermo-physical properties of superfluid helium for stabilization against thermal disturbances and for heat transport. The superconducting magnets operate in baths of "pressurized" helium II—i.e. above saturation pressure, in fact sub-cooled liquid—close to atmospheric pressure. The large, but finite thermal conductivity of helium II for heat fluxes of technical interest ensures that each bath is quasi-isothermal, but is however insufficient to transport heat over long distances in the accelerator tunnel. The 106.9 m long magnet strings are therefore cooled via a heat exchanger tube threading its way in the cold mass, inside which a small two-phase flow of saturated helium II gradually vaporizes as it absorbs heat from the pressurized helium II baths (Fig. 3.3). This efficient cooling scheme presents several benefits [2], among which those of avoiding the need for circulation pumps and of minimizing the need for flow space in the transverse cross-section of the magnet. The large size of the collider ring, and its implantation some 100 m underground, limiting access to eight points around its perimeter, result in sector lengths of 3.3 km individually serviced as concerns accelerator and technical systems.

**Fig. 3.2** The LHC cryomagnets installed in the tunnel (CERN photo)

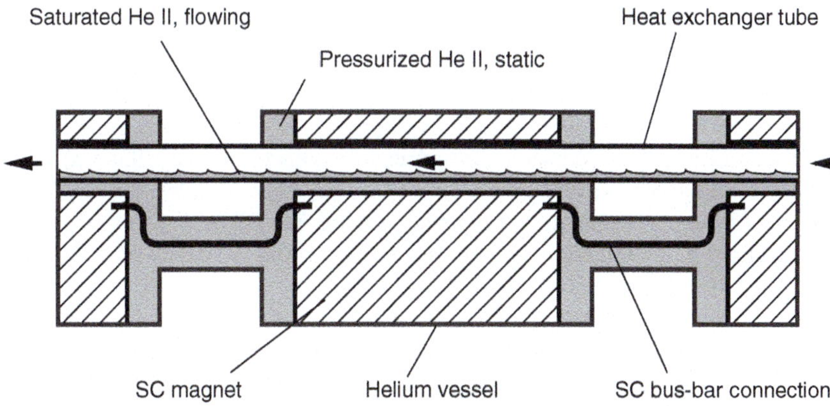


**Fig. 3.3** Principle of the superfluid helium cooling scheme of the LHC magnets

The LHC ring is cooled by eight large-capacity cryogenic plants, each normally cooling a 3.3 km sector, but able to serve two adjacent sectors and thus provide redundancy at partial load. This requires distributing and recovering helium flows at different temperatures in the access shafts and along the machine tunnel over lengths of up to 3.3 km, by means of a compound cryogenic distribution line running along the magnets and feeding each 106.9 m string in parallel. Each cryogenic plant provides a mix of liquefaction and refrigeration duties at 50–75 K and 4.5–20 K, amounting to an equivalent entropic capacity of 18 kW at 4.5 K. It is complemented at its cold end by a 1.8 K stage producing up to 2.4 kW of refrigeration power, by means of multi-stage sub-atmospheric compression using cold hydrodynamic and room-temperature volumetric compressors. Precooling the 37,500 t cold mass down to about 100 K is achieved by vaporization of some 10,000 t liquid nitrogen. Liquid nitrogen is otherwise not used in the system. 21,300 sensors monitor the machine cryogenics, while 4700 analog control loops ensure its operation. The cryogenic system of the LHC is described in Ref. [3].

In the following we tell the LHC cryostat story, approximately following chronological order, from the first studies and experimental work which enabled to establish the feasibility of such a large, novel superfluid helium system, to the several generations of prototypes meeting the evolution of machine configuration, their industrialization, series production and assembly, installation and interconnection in the tunnel, commissioning and operational experience. Focus is on the eight, 2818 m long continuous cold strings of the LHC arcs, containing in particular the 1232 dipole magnet cryostats (each 15 m in length) which constitute 85 % of the cold length and therefore drive the design, construction and industrialization choices, as well as the overall performance and cost of the system.

## 3.2 Feasibility of a Large Distributed Superfluid Helium System

The first approach to a large hadron collider to be installed in the existing tunnel housing the LEP collider, was based on twin-aperture dipole magnets using “cos θ” coils made of $Nb_3Sn$ superconductor operated in baths of normal helium at 4.5 K, to produce a bending field around 10 T [4]. The study included the conceptual design of the corresponding cryogenic system, including cryostat simplified cross-section. It identified the main requirements and boundary conditions of large-capacity refrigerators at 4.5 K, thermal shielding at around 80 K, cryogenic fluid distribution along the 3.3 km sectors, cryogen storage, cooldown of the large magnet mass, quench handling and recovery and underground safety.

Soon after this it became clear that, in view of the limited availability, high cost and technical difficulty of implementing $Nb_3Sn$ superconductors in accelerator magnets, the alternative of operating the more conventional Nb-Ti superconducting alloy at lower temperature in superfluid helium—to boost its current-carrying capacity at high field—should also be explored. This technique, pioneered at CEA Grenoble and first applied to cool high-field magnets for condensed-matter physics (the Grenoble High-Field Magnet) and nuclear fusion research (the TORE SUPRA tokamak at CEA Cadarache) [5], had however never been applied to the magnets and distributed configuration of a large circular accelerator. It clearly appeared that the prime feasibility issues to be investigated were the management of heat loads (and thus the design of a cryostat meeting the required thermal performance) and the cryogenic distribution scheme—the latter also impacting the cryostat design to incorporate the cryogenic pipelines. A preliminary study of a superfluid helium cryogenic system for the LHC was then conducted [6].

A novel feature of the LHC with respect to previous superconducting accelerators is that the "static" heat loads, i.e. the heat in-leaks to the cold mass and other cryogenic components of the machine, represent only a fraction of the total. The large excitation currents of the magnets—12 kA for the LHC dipoles—induce resistive heating in the numerous non-superconducting joints between the winding layers, between the coils and between the magnets in a string: this additional "dynamic" heat load necessarily ends up into the helium bath at the lowest temperature level and must therefore be contained by a tight specification on the maximum joint resistance (typically lower than 1 nΩ per joint). Moreover, the circulation of intense particle beams in the accelerator rings also induces dynamic heat loads, through several processes such as dissipation of beam image currents in the resistive wall of the beam pipes, synchrotron radiation from the beams (in the UV range for LHC protons) and particle losses from the beam halo. While the latter produce cascades of secondary particles eventually depositing their energy into the magnet cold mass, the power of the other two processes may be intercepted at higher temperature in order to reduce its thermodynamic impact on the refrigeration system. This is the primary function devoted to the beam screens, the concept of which was introduced for the first time in the study. Today, beam screens cooled by supercritical helium between 5 and 20 K equip the LHC magnets and limit the heat load to the 1.9 K level to acceptable values; they also perform several other essential functions [7], in particular that of a distributed cryopump maintaining a good level of dynamic vacuum in the beam pipes, and their elaborate design is therefore the optimized result of a complex process involving several domains of accelerator physics and technology (Fig. 3.4). Pre-design of the cryostat featuring a 90 K thermal shield, non-metallic supports of the cold mass with heat intercepts at 90 and 4.5 K, multilayer reflective insulation around the cold mass, and integrating cryogenic distribution pipelines, gave an estimated heat in-leak at 1.9 K of 0.15 W/m. Containing the dissipation in the resistive joints to 0.05 W/m and introducing the beam screens to intercept most of the beam-induced heat loads yielded a total steady-state budget of 0.2 W/m at 1.9 K, to be compared with 0.9 W/m at 4.5 K for a normal helium cryostat in absence of beam screen. The conclusion of this preliminary study was that a superfluid helium cryogenic system for the LHC, including magnet cryostat and distribution scheme, appeared possible, though more complex than one at normal helium temperature.

The following years saw the refinement of such studies, particularly concerning the cryogenic distribution scheme and the assessment of dynamic heat loads, but the cryostat design concept (Fig. 3.5) and estimated performance of 0.15 W/m at 1.9 K were essentially confirmed [8]. The addition of a complete shield at 5 K surrounding the cold mass allowed—in principle—to reduce the residual gas conduction to the cold mass and thus the dependence of total heat load on the quality of

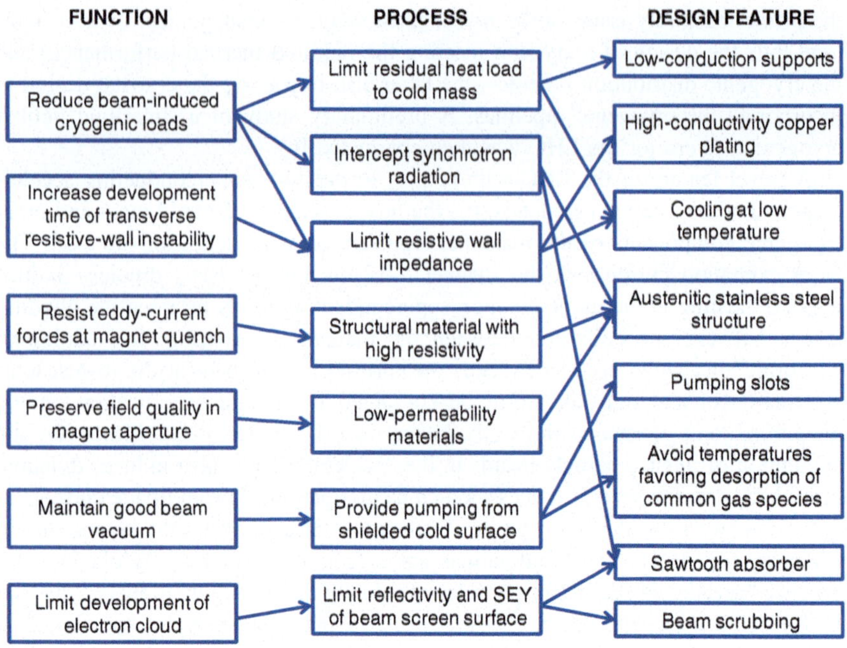


**Fig. 3.4** Functional design map of beam screen for particle accelerator

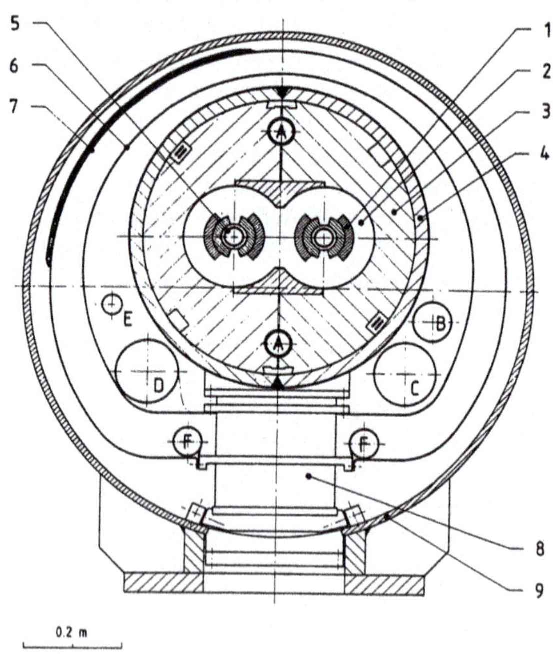


**Fig. 3.5** Transverse cross-section of conceptual cryostat housing LHC twin-aperture dipole magnet at 1.9 K. *1* magnet coils; *2* magnet collars; *3* magnet yoke; *4* magnet shrinking cylinder and helium vessel; *5* beam screen; *6* shield at 5 K; *7* MLI insulated shield at 80 K; *8* non-metallic support post; *9* vacuum vessel; *A* helium II heat exchanger tube; *B–F* cryogenic distribution and recovery lines at 1.9, 5 and 80 K

the insulation vacuum. Assessment of the practical benefit of this additional feature led to several iterations discussed in the following.

The early conceptual studies of LHC cryostats benefitted, at least down to normal helium temperature, from the extensive design and development work performed in the USA, particularly at Fermi National Accelerator Laboratory in preparation for the SSC project [9]. Still, experimental validation of the LHC cryostat design proper was required, first by measurement of critical components and eventually on a full-scale prototype. A dedicated test set-up was built to perform precision heat in-leak measurements at 80, 4.2 and 1.8 K in the laboratory, combining several independent methods (cryogen boil-off, "heat-meters" i.e. calibrated thermal impedances, and transient temperature rise in a superfluid helium bath). Enabling the measurement of heat loads down to the mW range at 1.8 K, this experimental set-up was used *inter alia* to qualify support posts made of non-metallic composite materials (glass-fibre/epoxy and carbon-fibre/epoxy) and to optimize position and practical realization of their heat intercepts [10]. Another critical source of heat in-leak in the LHC cryostat stems from the 50,000 $m^2$ lateral surface area of the cold mass—the equivalent of seven soccer fields—receiving heat by radiation and residual gas conduction from the 80 K thermal shield in case a 5 K shield is not used or difficult to thermalize at its nominal temperature. It was therefore necessary to compare, on samples of practical geometry and size, the performance of several reflective single-layer and multilayer systems at low boundary temperature and in different conditions of insulation vacuum. This was done on a dedicated set-up housing a 3 $m^2$ cylindrical sample of the insulation system to be tested, with possibility of varying the warm and cold boundary temperatures [11]. It confirmed that very low heat fluxes can be achieved by passive, reflective systems in good vacuum, and that multilayer systems show a clear advantage to limit heat flux runaway in case of degraded insulation vacuum.

An opportunity to build rapidly a quasi-full-scale, twin-aperture prototype dipole was provided by the termination of series production of superconducting magnets for the HERA proton ring. Two sets of spare Nb-Ti dipole coils from the HERA production line were integrated into a common yoke to build a 9.15 m long twin-aperture prototype (TAP), to be operated in superfluid helium in order to simulate a LHC dipole: it was estimated that a field of 7.5 T could be produced with the coils powered at 8610 A. The TAP magnet needed a dedicated cryostat to be operated and tested in superfluid helium. Conversely, this provided a unique opportunity for experimenting, for the first time and almost in full scale, the design of the LHC cryostats and for studying some of the thermal and mechanical problems associated with their cryogenic operation.

The TAP cryostat [12] was designed at CERN along the principles of the conceptual LHC cryostats (Fig. 3.6) and built by industry. In addition to the magnet, it housed cryogenic distribution pipelines—some of which were used for heat interception—and featured three glassfibre/epoxy posts supporting the magnet, the central one fixed longitudinally and the other two allowed to follow the thermal contraction of the cold mass via rollers on tracks fitted to the vacuum vessel. These rollers and tracks were also used during assembly to insert the cold mass into the

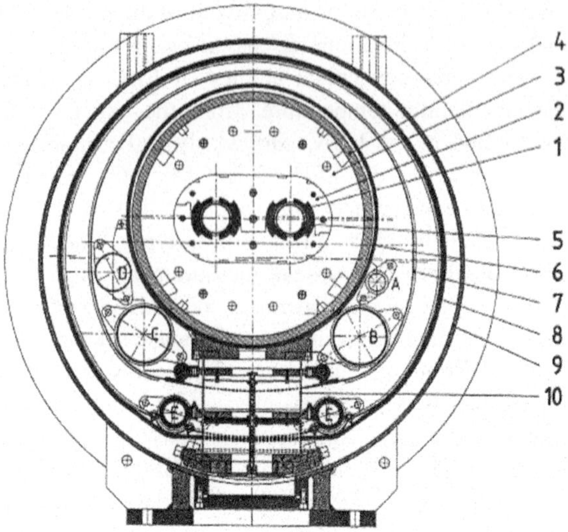


**Fig. 3.6** Transverse cross-section of the TAP magnet and cryostat. *1* HERA-type superconducting coils; *2* magnet collars; *3* magnet yoke; *4* magnet shrinking cylinder; *5* cold bore tube; *6* helium vessel; *7* screen at 5 K; *8* thermal shield at 80 K with MLI; *9* vacuum vessel; *10* support post; *A & B* 1.8 K helium pipes; *C & D* 4.5 K helium pipes; *E* liquid nitrogen pipes

vacuum vessel. Demountable shipping restraints were provided to take transport acceleration loads. A liquid-nitrogen cooled thermal shield wrapped with 30 layers of MLI intercepted most of the radiative heat in-leak from room temperature, and a second screen surrounded the cold mass and pipeline assembly, actively cooled from 4.5 K pipes. The all-welded, austenitic stainless steel (AISI 304L) helium enclosure had a design pressure of 2 MPa to allow fast cooldown/warmup and resist resistive transitions of the magnet. The vacuum vessel was made of unalloyed construction steel (with austenitic stainless steel flanges), for reasons of magnetic shielding of the stray flux and cost in view of a large series production. Its inner surface was protected by low-outgassing epoxy paint compatible with high-vacuum operation. The cylindrical shell was provided with reinforcement rings at the longitudinal positions of the cold mass supports.

The TAP magnet was integrated into its cryostat by industry and shipped to CEA Saclay for cold tests. The magnet and cryostat performed very well, with maximum field of 8.25 T reached after seven resistive transitions, excellent leak-tightness of the superfluid helium enclosure, heat in-leak below 1 W at 1.8 K and about 4.6 W at 4.5 K [13]. Following resistive transitions of the magnet, high values of peak pressure, up to 3 MPa, were measured in the annular space between coils and cold bore tube at mid-length, thus raising concern about the choice of design pressure for the helium enclosure. A dedicated study based on a hydrodynamic model using thermo-physical properties of helium [14] confirmed the measurements and gave indications to limit this peak pressure, particularly when considering longer magnets. From the cryostat point of view, the TAP project established the soundness of the design principles and boundary conditions, confirmed performance expectations from design calculations and measurements on partial components and demonstrated the feasibility of industrial production at market price—including quality assurance, licencing procedures and operational safety.

## 3.3 Prototype Cryostats and String Tests

Based on the previous experience, a small series of cryostats were designed at CERN and procured from industry to house the first generation of 10 m long prototype dipole magnets, to be tested individually and later connected and operated in a string. The cryostats also had to accommodate the cryogenic distribution pipework sized for feeding a full LHC sector. Design requirements included heat in-leaks meeting the allowed thermal budget at the three temperature levels or ranges in the LHC cryogenic system (50–75 K, 4.5–20 K and 1.9 K), provision of a stable, precise and reproducible mechanical support for the magnet, withstanding of thermal and mechanical transients associated with rapid cooldown and magnet resistive transitions, as well as all possible failure modes. Technical solutions adapted to large series production were favoured.

The resulting design [15], inspired by the TAP cryostat but with a number of variants, is shown in Fig. 3.7. The cold mass support system still used three glassfibre/epoxy posts with two levels of heat interception, but the functions of assembly and of accommodation of longitudinal contractions were now handled separately, the former provided by a removable assembly trolley and the latter by sliding plates incorporating low-friction material. The support posts proper were developed and their construction optimized based on mechanical and thermal performance measurements [16]. The thermal shield at 80 K, made in commercially pure aluminium (AA 1100) to ensure good azimuthal temperature uniformity, consisted of a rigid lower tray incorporating a welded extruded aluminium profile providing the cooling channel, onto which a roughly cylindrical upper shell was tack-welded in four longitudinal sections. This compound construction provided sufficient mechanical rigidity and azimuthal thermal conduction to limit deflections of the shield during thermal transients, and thus avoid damage or thermal shorts.

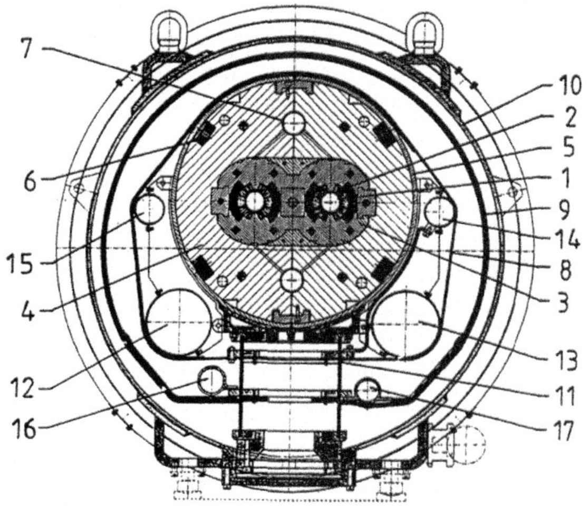


**Fig. 3.7** Transverse cross-section of first-generation prototype dipole and cryostat. *1* cold bore; *2* magnet coils; *3* magnet collars; *4* magnet yoke; *5* magnet shrinking cylinder and helium vessel; *6* bus bars; *7* He II heat exchanger bore; *8* cold mass multilayer insulation; *9* thermal shield with multilayer insulation; *10* vacuum vessel; *11* support post; *12* 1.8 K GHe line; *13* 10 K GHe line; *14* 4.5 K He line; *15* 2.2 K LHe line; *16* 50 K GHe supply; *17* 70 K GHe return

The actively cooled 4.5 K screen surrounding the cold mass of the TAP cryostat was replaced here by a 10-layer blanket of MLI, partially thermalized onto the 4.5 K lines but insulated from the cold mass and 2 K lines by polyester spacers. Another, important novelty with respect to the TAP cryostat was the first engineering study and optimization of magnet and cryostat interconnection. To maximize the magnet filling factor of the machine and thus the beam energy reach, the interconnection space must be kept as short as possible, while allowing to reliably perform, inspect and test some 14 orbital welds joining 11 pipes. Space for automatic orbital welding and cutting machines, as well as for "clam shells" for local leak-testing of welds, was taken into account. The interconnection also includes all bellows accommodating construction and alignment tolerances, as well differential thermal contractions and displacements occurring in all possible phases of operation, including failure modes (as a reminder, the LHC cold mass shrinks by some 80 m in perimeter upon cooldown). Bellows specifications, including travel, misalignments, maximum rigidity and fatigue lifetime were established, pointing to hydro-formed multiply construction with external shells against buckling where needed. Samples were procured from several bellows manufacturers and extensively tested, thus providing a qualified industrial basis for future series procurement.

The design of the prototype dipole cryostat was adapted to house the main quadrupole, chromaticity sextupole and corrector magnets, beam position monitors and local cryogenic equipment as well as a vacuum barrier segmenting the insulation vacuum, all included in a so-called short-straight-section cryostat [17], and a prototype was built and tested.

In parallel with these activities, preparation was made to perform cryogenic tests of the magnets and cryostats. A 7200 $m^2$ test hall already equipped with a 6 kW at 4.5 K cryogenic refrigerator was further provided with a low-pressure helium pumping unit capable of 6 g/s at 1 kPa, upgradeable to 18 g/s with the addition of a cold booster compressor. Cooldown and warmup units, each providing up to 120 kW down to 100 K by vaporization of liquid nitrogen [18] were installed. The test station was planned to develop in a modular way following the ramp-up of series production, with up to twelve parallel test benches, the first two of which were commissioned to test prototypes [19]. Each bench, equipped with 18 kA current leads, could be operated independently of the others, only sharing—when and as much as needed—the common facilities such as normal liquid helium supply, pumping capacity at 1 kPa and gaseous helium flow for cooldown and warmup.

At the same time it was noted that precision thermal measurements on cryostats mounted on magnet test benches would be difficult to perform, due to parasitic heat loads (in particular the comparatively high heat in-leak coming from the anti-cryostats installed in the magnet apertures to perform magnetic measurements) as well as to schedule interference (reaching thermal steady-state requires long stabilization time incompatible with magnet turnaround on test benches). It was therefore decided to build a full-scale thermal model [20] from a 10 m prototype cryostat equipped with a dummy cold mass and finely instrumented. The cryostat thermal model could be operated independently of the magnet tests station, allowing to vary parameters such as insulation vacuum and temperature levels, and providing precision of ±0.03 W at 1.8 K, ±0.1 W at 5–20 K and ±3 W at 50–75 K. Heat load measurements in nominal operating conditions on the thermal model showed excellent agreement with calculated values [21]. Moreover, several off-design runs enabled to measure the sensitivity of the cryostat heat in-leaks upon the temperature of heat intercepts and thermal shield, as well as on the residual pressure in the vacuum insulation space, also in good agreement with the calculations. The thermal performance of the prototype cryostats was therefore well understood in nominal and quasi-nominal conditions, and a mathematical model benchmarked on the experimental measurements could be used reliably to quantitatively assess its degradation in off-design conditions [22]. The pending question of the inner screen surrounding the cold mass, either "floating" or tentatively thermalized on the 5 K line, was then revisited [23], using realistic values of impedance for the thermalization contacts and net-type insulating spacers in MLI: the study indicated potential improvement for a thermalized system, from 60 $mW/m^2$ down to 30 $mW/m^2$ in good insulation vacuum ($10^{-4}$ Pa), at the cost of increased complexity in cryostat assembly.

Another pending issue was that of using normal construction steel for the cryostat outer vessel, normally operating at room temperature but possibly subject to being cooled below the ductile-to-brittle transition temperature in case of catastrophic loss of insulation vacuum. This was tested on a full scale cryostat equipped with a dummy cold mass initially cooled at 80 K [24]. Upon breach of the vacuum with gaseous helium, heat transfer to the cold mass induced water condensation and frost formation on the outer vessel, but its minimum temperature reached at 260 K remained above the acceptable limit. Moreover, temperature evolution was slow enough to allow detection of the accident and implementation of remedial action, e.g. by bringing additional pumping speed to restore, at least partially, the insulation vacuum.

The prototype magnets in their cryostats were then assembled stepwise into a test string [25], a full-scale working model of the elementary cell of the machine lattice and local cryogenics, installed on a 1.4 % ramp corresponding to the maximum slope of the LHC tunnel, with a final length of 50 m [26]. The test string (Fig. 3.8) enabled us to practice assembly, leak-tightness and alignment procedures, then mechanical stability and vacuum measurements upon controlled cooldown of the 65-t cold mass, and finally cryogenic operation and control, magnet powering and discharge, handling of resistive transitions and recovery [27]. Vacuum and thermal transients were investigated, in particular as concerns non-standard operation and failure modes [28]. In spite of the absence of beam, the test string also served as an excellent training ground for young recruits, who eventually had to operate from a dedicated control room a compound system integrating several accelerator technologies, with many measurements and control loops. Finally, it constituted an important demonstrator which contributed to the approval of LHC construction by the CERN Council.

**Fig. 3.8** First-generation LHC Test String (CERN photo)

In the last round of studies before project approval, several important changes were implemented in the configuration of the LHC and its cryogenic system. To minimize their number and the associated manufacturing and testing costs, the LHC dipole magnets had their length extended from the previous 10 m to the maximum which could be practically transported in European roads (40-t limit) and conveniently handled for assembly, testing and installation, i.e. 15 m, thus requiring new, longer cryostats. In order not to lose in useful aperture, the cold mass had to be curved, with a horizontal sagitta of 9 mm, but the cryostats could remain straight. As concerns cryogenic system architecture, grouping all refrigeration equipment in the five points around the ring (instead of eight) already equipped with technical infrastructure led to significant cost savings, and allowed for redundancy at part-load, but cryogenic flows had then to be distributed over the full 3.3 km length of the sectors (instead of half-sectors). This required larger-diameter pipes which became difficult to integrate in the cryostat transverse cross-section. Consequently, all distribution pipework and local equipment was moved to a compound cryogenic line running in the tunnel along the magnets, and the remaining pipework in the cryostats was then limited to that needed for operation of the cryostat proper. Another important consequence is that the cryogenic line could be installed, commissioned and tested independently of the progress in magnet installation, thus bringing valuable schedule flexibility in the last phase of project construction [29]. Moreover, it had been observed that in view of the large size of the LHC ring, "passive" cryogenics, i.e. magnet cryostats and cryogenic distribution represented the largest share of the cost breakdown, in fact larger than "active" cryogenics, i.e. the refrigeration plants. Consequently, any simplification of the cryostats and cryogenic distribution scheme, even though it may result in increasing the refrigeration duty, would concur to lowering the total cost. A simplified cryogenic scheme was then implemented [30], with local cooling loops extending over the length of a full lattice cell (106.9 m) instead of the previous half-cell, implementation of local 4.5–2.2 K sub-cooling heat exchangers in each local cooling loop allowing the suppression of one pipe inside the distribution line, and a large overall reduction in the number of cryogenic components.

A second generation of prototype cryostats for dipoles [31] and short-straight-sections [32] was then designed and built, based on the experience acquired with the first-generation prototypes and incorporating the changes in LHC configuration described above. In the dipole cryostat (Fig. 3.9) the smaller outer diameter (914 mm) allowed the shell of the vacuum vessel to be made of standard, helically welded elements for pipelines, with a wall thickness of 12 mm. To prevent any risk of embrittlement, the construction steel used (DIN GS-21 Mn5) was however specified to have a minimum energy absorption of 28 J/cm$^2$ as measured in an ISO-standard Charpy test at 240 K. The shells were equipped with AISI 304L end flanges and reinforced, at the longitudinal positions of the magnet supports, by welded rings and forged brackets for lifting points and alignment fiducials. The longer, heavier cold mass was still supported on three posts, the longitudinal positions of which were compromised between minimizing vertical sagitta (Bessel conditions) and keeping the ends of the cold mass horizontal to ease interconnect assembly (Airy conditions). The vertical off-axis between magnet and cryostat axes was reduced to 80 mm, thus requiring shorter support post of 210 mm total height, redesigned to take higher combined loading of 125 kN in axial compression and 40 kN in shear [33]. The longer thermal shield, a quasi-cylinder cooled from one

**Fig. 3.9** Second-generation prototype cryostats assembled into second Test String (CERN photo)

generatrix, was prone to larger deformations upon thermal transients, with the risk of damage and unwanted thermal shorts [34]. The technical solution experimented in the first-generation prototypes, i.e. separating the functions of mechanical rigidity and azimuthal temperature homogeneity, was further refined towards industrialization: the lower tray was made of a full-length extrusion of A6060 aluminium

**Fig. 3.10** Extruded aluminium alloy (A6060) bottom tray of cryostat thermal shield (CERN photo)

alloy, integrating two 80 mm diameter channels, one used for the cooling, the other left open (Fig. 3.10). The cooling channel was fitted at the ends with aluminium-to-stainless steel pipe transitions for interconnection. The upper shell of the thermal shield made of commercially pure aluminium A1100, was segmented in five sections (Fig. 3.11). A semi-rigid screen, thermalized to the 5 K pipes providing heat intercepts on the support posts was tentatively re-introduced in these cryostats with the aim of further reducing the heat in-leak on the cold mass. The cryostat interconnect was also redesigned to accommodate installation misalignments of up to ±4 mm in any direction and thermal contractions of up to 48 mm between the longitudinally fixed central posts of neighbouring dipoles. The short-straight section cryostat (Fig. 3.12), designed in collaboration with IPN Orsay (France) incorporated many design features of the dipole cryostat, and showed a similar transverse cross-section. It was equipped with a vacuum barrier for segmenting the insulation vacuum, and a cryogenic service module connected to the cryogenic distribution line by a multi-pipe "jumper" connection. The service module also contained beam position monitors, conduction-cooled current leads for the independently-powered orbit corrector magnets, and instrumentation feedthroughs. However, all local cryogenic equipment was now moved to the cryogenic line on the other side of the jumper connection.

Fig. 3.11 Aluminium A1100 shell of cryostat thermal shield (CERN photo)

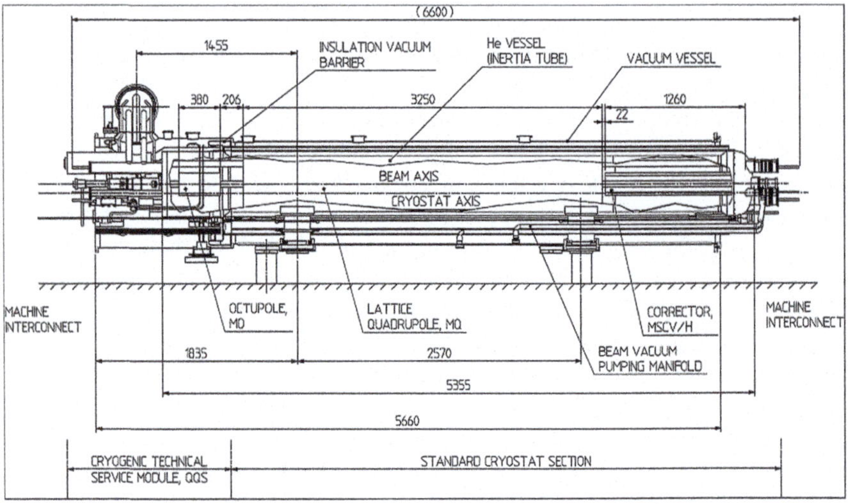


**Fig. 3.12** Longitudinal cross-section of the second-generation short-straight-section cryostat

Five second-generation dipole cryostats, and two short-straight section cryostats were procured from industry and assembled around their magnets for cold testing and installation into a 107 m long second-generation test string [35], the last large-scale experimental validation of the LHC accelerator systems before entering construction. A first systematic risk analysis of the LHC cryogenic system [36] established all possible failure modes and analysed their consequences, thus confirming the cryostat design parameters and safety features.

## 3.4 Industrial Series Production, Installation and Commissioning

The design of the series cryostats (Fig. 3.13) derived directly from the second-generation prototypes, the measured thermal performance of which was used to establish reliably a reference thermal budget for the purpose of sizing the cryogenic refrigeration plants and distribution system. A "heat-load working group" involving all stakeholders was established to collect all calculated and measured data, validate the estimates (Table 3.1) and track possible future evolutions.

Changes of a few manufacturing aspects streamlined the design towards industrialization in view of series production [37]. Trading moderate improvement

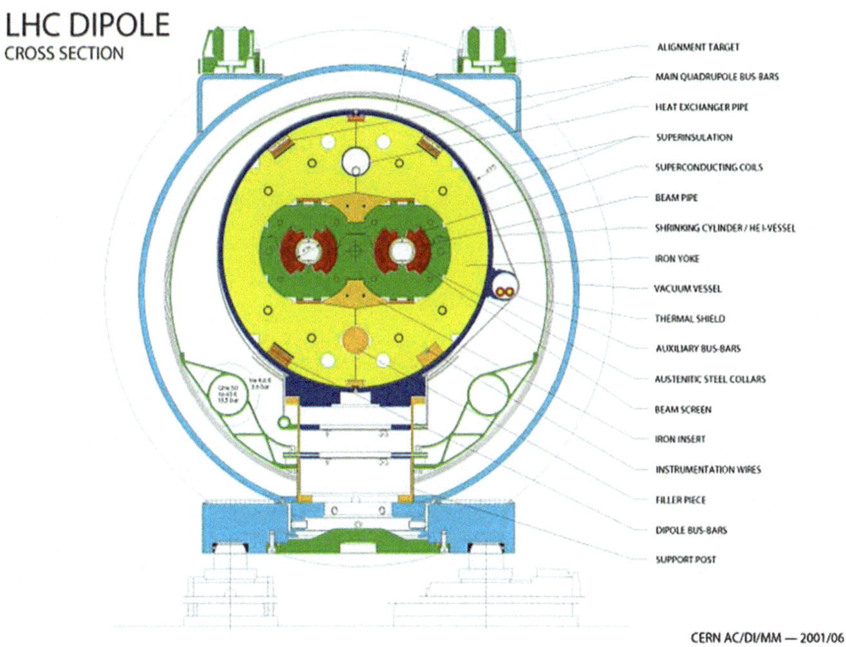


**Fig. 3.13** Transverse cross-section of series dipole cryostat

**Table 3.1** Heat in-leaks of 106.9 m long standard cell (6 dipoles and 2 short straight sections)

| | 50–75 K | 4.5–20 K | 1.9 K |
|---|---|---|---|
| Magnet support posts (W) | 157 | 9.78 | 1.07 |
| Thermal shield (W) | 301 | – | – |
| Cold mass MLI (W) | – | 0.85 | 11.3 |
| Beam screen (W) | – | – | 1.62 |
| Instrumentation feedthrough (W) | – | – | 4.25 |
| Beam vacuum feedthrough (W) | 2.40 | – | 0.42 |
| Orbit corrector current leads (W) | 10.4 | 2.44 | 0.53 |
| Beam position monitor (W) | – | 0.93 | 0.60 |
| Cryogenic service module (W) | – | 0.01 | 0.21 |
| Vacuum barrier (W) | 11.6 | 0.03 | 0.42 |
| **Total heat in-leak (W)** | **482** | **14.0** | **20.4** |
| **Average linear heat in-leak (W/m)** | **4.51** | **0.13** | **0.19** |

**Fig. 3.14** Glassfibre-reinforced epoxy support post (CERN photo)

of thermal performance against complexity and cost, the 5 K thermalization of the screen was finally not retained, and the cold mass was eventually covered with a 10-layer blanket of MLI. The complete rationale for this choice is summarized in Ref. [38]. Following completion of welding work, the series vacuum vessels were stress-relieved by vibration to ensure mechanical stability over time. Instead of the previously used low-outgassing epoxy paint, their inner surface was sandblasted and simply cleaned to high-vacuum standards, with no other surface treatment. Magnet support posts (Fig. 3.14) were composed of monolithic 4 mm thick columns of glass-fibre reinforced epoxy, manufactured by resin-transfer moulding, a highly automated technique ensuring low production cost and reproducible quality for series of 4700 units [39]. At the heat intercept locations, external flanges of aluminium alloy and internal stainless-steel reinforcement rings are glued to the columns. Aluminium anti-radiation disks and reflective coating on the lower, warmer section exposed to radiation from the room-temperature outer vessel, complete the assembly. The 30-layer MLI around the thermal shield and 10-layer MLI around the cold mass were made of pre-fabricated blankets of 6 mm thick PET film, double-aluminized with 400 Å, interleaved by a very low weight polyester spacer. The developed perimeter of each blanket was adjusted to compensate for their differential thermal contraction with the stainless-steel and aluminium shells supporting them. Besides more precise manufacturing, pre-fabrication enabled the multilayer insulation systems to be assembled by non-specialized personnel, using Velcro™ fasteners, and avoided installation errors, thus ensuring future thermal performance of the multilayer insulation systems in the LHC ring.

For series production, it was decided to get the cryostats assembled by industry on the CERN premises, with CERN procuring all components and providing the contractors with a build-to-print specification, the use of two assembly halls totalling some 10,000 $m^2$ floor space, lifting and transport facilities, heavy and specialized tooling, detailed procedures and quality management tools, as well as with initial training of the execution personnel on the pre-series units, co-assembled with CERN experts (Figs. 3.15 and 3.16). Assembly of the 1232 dipole and 474 short-straight-section cryostats spanned five years, representing some 850,000 h of work with personnel numbers—operators and technicians—peaking at 145 [40].

**Fig. 3.15** Dipole cryostat assembly hall (CERN photo)

**Fig. 3.16** Short-straight-section cryostat assembly hall (CERN photo)

Following assembly, each cryo-magnet was moved to the test hall using a specially designed vehicle used for both lifting and transport (Fig. 3.17), connected to one of the twelve test benches, pumped down, cooled down and cryogenically tested, with the magnet powered up to 10 % above nominal (Fig. 3.18). Magnetic measurements in cold conditions were only performed on a fraction of the production. After removal from the test bench each cryo-magnet was allocated a position in the ring based on test results, prepared for installation and lowered to the tunnel level via a large, elliptical shaft equipped with a 40-t gantry crane with a hook travel of 54 m. The major axis of 18 m of the elliptical shaft is able to accommodate the full assembled length of a cryo-dipole, with protection end covers, horizontally (Fig. 3.19).

Cryostat assembly and cryogenic tests followed the production rate of magnets in industry (Fig. 3.20), while installation in the tunnel started later and proceeded at a faster rate. As a consequence, a large buffer storage of assembled and tested cryo-magnets was available, enabling the optimization of their final installation positions in the accelerator tunnel in order to reduce dispersion in the magnetic and geometrical properties of assembled sectors. The cryo-magnets were stored in the open air for several months, with their ends capped and insulation space kept under dry nitrogen.

**Fig. 3.17** Completed dipole cryo-magnet transported to test hall (CERN photo)

**Fig. 3.18** Cryogenic tests of magnets at CERN (CERN photo)

Transport in the tunnel from the shaft to the installation position was done via an optically guided, electrically powered trolley at low velocity (3 km/h) to minimize inertial forces in view of the large masses transported. On location, the cryo-magnet weight was loaded to a transverse transfer platform, and then transferred onto its pre-positioned final support jacks (Fig. 3.21). Electrical magnet interconnections and hydraulic cryostat interconnections were then performed. This was done by

**Fig. 3.19** Lowering of cryo-magnet to LHC tunnel via elliptical shaft (CERN photo)

industry in the framework of a build-to-print contract. CERN developed the methods, special tooling and quality assurance procedures to be used by the contractor, and provided initial training of its personnel [41]. Some 40,000 cryogenic pipe welds were reliably executed using automatic TIG orbital welders (Fig. 3.22). The inner part of a completed cryo-magnet interconnection, with the MLI blankets, thermal shield and outer vacuum sleeve removed, is shown in Fig. 3.23. The domed ends of the helium vessels, as well as the reflective multilayer insulation blankets covering the cold mass and the thermal shield, are clearly visible on either side of the interconnection.

Following global pressure and leak-tightness tests, each 3.3 km sector of the LHC was cooled down and powered. Total sector heat in-leaks (Fig. 3.24) were measured within the estimated values [42], a final validation of the quality of the design, construction and installation of the cryostats.

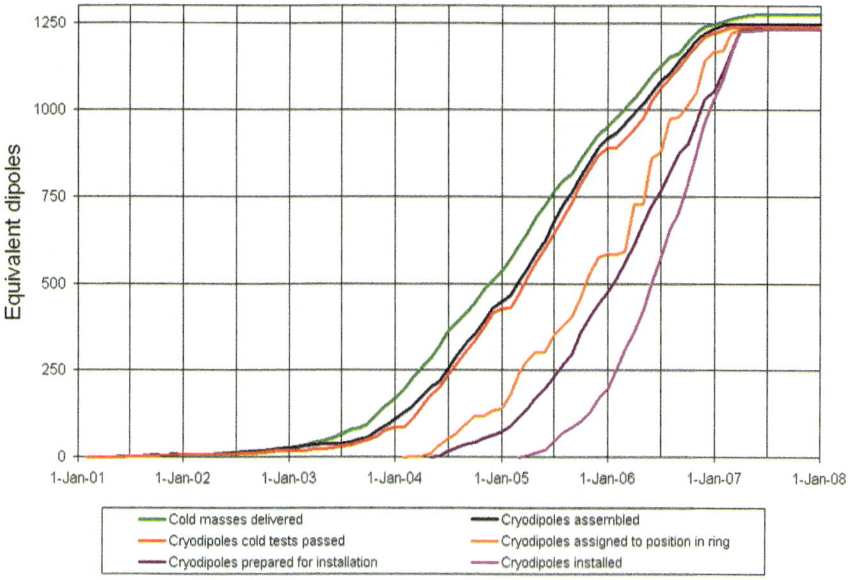


**Fig. 3.20** Series production of LHC dipole cryo-magnets

**Fig. 3.21** Transverse positioning of cryo-magnet on final location (CERN photo)

**Fig. 3.22** Cryogenic piping interconnection in LHC tunnel by orbital TIG welding (CERN photo)

**Fig. 3.23** Cryostat interconnection inner piping (CERN photo)

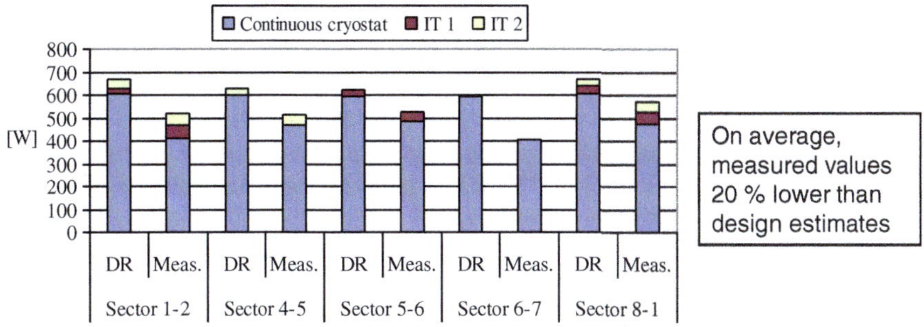


**Fig. 3.24** Measured heat in-leaks of LHC sectors at 1.9 K (*DR* design report estimates; *IT* inner quadrupole triplet)

## 3.5 Concluding Remarks

The LHC contains the largest system of advanced, high-performance cryostats housing superconducting accelerator devices, and may thus be considered, *mutatis mutandis*, as reference for future, similar projects. Several lessons can be drawn from this experience.

The technology of superfluid helium can be reliably applied to very large systems built by industry. In particular, the design principles, construction techniques and quality assurance methods used in normal helium cryostats are fully adequate to ensure leak-tightness of superfluid helium enclosures.

Cryostats of superconducting accelerator devices, magnets or RF cavities, must meet complex and often conflicting geometrical, mechanical and thermal requirements, at the interface between accelerator physics and technology requirements, industrial manufacturing constraints and cost. Their design must therefore follow closely the changes in configuration of the project and sometimes drive them. It is therefore essential for the cryostat designers to be fully integrated with the project team and in close contact with the magnet, RF, cryogenics and vacuum groups.

Within the cryogenic system, the total cost of the cryostats and distribution lines often exceeds that of the refrigeration plants. Global optimization of the complete system is therefore necessary, which may lead to simplify cryostat construction at the cost of loss in thermal performance, rather than designing for lowest possible heat-in-leak.

The large amount of thorough design and development work performed at CERN and in partner laboratories on the LHC cryostats has contributed to significantly increase knowledge and confidence in the design, engineering and quality assurance methods for this type of equipment. It is interesting to note that this work has already found an application in another project. Cryostats similar to those of the LHC dipoles are housing the superconducting magnets of the beam line for the T2K long-baseline neutrino experiment at the J-PARC laboratory in Tokai, Japan [43]. On the training and education side, a tutorial on cryostat design inspired from the development of LHC cryostats was first prepared for the CERN Accelerator School

39. V. Parma et al., The LHC cryomagnet supports in glass-fibre reinforced epoxy: a large-scale industrial production with high reproducibility in performance. Adv. Cryo. Eng. **54**, AIP Conf. Proc. **986**, 211–218 (2008)
40. A. Poncet, V. Parma, Series-produced helium II cryostats for the LHC magnets: technical choices, industrialization, costs. Adv. Cryo. Eng. **53A**, AIP Conf. Proc. **985**, 739–746 (2008)
41. J-Ph Tock et al., The interconnections of the LHC cryomagnets at CERN: strategy applied and first results of the industrialization process. IEEE Trans. Appl. Supercond. **18**(2), 116–120 (2008)
42. S. Claudet at al., *Cryogenic heat load and refrigeration capacity management at the Large Hadron Collider*, in Proceedings of ICEC22—ICMC 2008, KIASC (2009), pp. 835–840
43. T. Nakamoto et al., Construction of superconducting magnet system for the J-PARC neutrino beam line. IEEE Trans. Appl. Supercond. **20**(3), 208–213 (2010)
44. P. Lebrun, *Design of a cryostat for superconducting accelerator magnets: the LHC main dipole case*, in Proceedings of CAS Superconductivity and Cryogenics for Accelerator and Detectors, CERN-2004-008, ISBN 92-9083-230-4 (2004), pp. 348–362
45. V. Parma, *Cryostat design*, in Proceedings of CAS Superconductivity for Accelerators, CERN-2014-005 (2014), pp. 353–399. ISBN 978-92-9083-405-2

# Chapter 4
# The Superfluid Helium On-Orbit Transfer (SHOOT) Flight Demonstration

**Michael DiPirro**

**Abstract** Cryostats for space applications have unique requirements including significant weight and volume limits as well as the need to take into account zero gravity when designing fluid systems. This chapter describes the design, testing and space operations of the Superfluid Helium On-Orbit Transfer (SHOOT) flight demonstration. Topics covered in the design of the SHOOT cryostats include: structural and thermal insulation design, phase separation and liquid acquisition systems, the use of thermomechanical pumps, instrumentation and safety. Operations on both the ground and on orbit are also discussed.

## 4.1 Introduction

In the 1980s NASA recognized that there would be an increasing demand for cryogenic instruments in space. In particular, cooling to below 4 K would become routinely required for advanced astronomical instruments. In that period cryocoolers for this temperature range were in their infancy; none provided long life, high reliability, and low input power required for space flight. This left stored cryogens, uniquely liquid helium, to be required for many future missions. Refueling helium dewars in orbit was being studied to extend these instrument's lifetimes in the early days of the space shuttle and in the concept phase of the International Space Station [1]. The Superfluid Helium On-Orbit Transfer (SHOOT) Flight Demonstration was proposed in 1984 to carry out a series of tests of the technology and physics required to store and move liquid helium from a large depot into individual dewars aboard various satellites.

The experimental components for SHOOT, which drove the requirement for an on-orbit test, were the liquid acquisition devices (LADs). These LADs could only be thoroughly tested in the proper acceleration conditions, i.e., in a very low acceleration for an extended period of time, in orbit. Of course one must demon-

M. DiPirro (✉)
Code 552, NASA Goddard Space Flight Center, Greenbelt, MD 20771, USA
e-mail: mike.dipirro@nasa.gov

 
J.G. Weisend II (ed.), *Cryostat Design*, International Cryogenics Monograph Series,
DOI 10.1007/978-3-319-31150-0_4

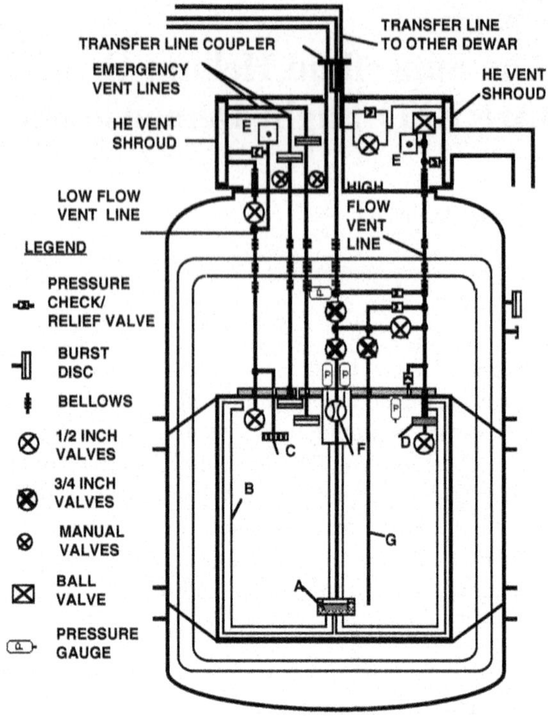


**Fig. 4.1** Schematic of one of the SHOOT dewars. *A* TM pump; *B* liquid acquisition device; *C* HeI/HeII phase separator (also known as the low-flow phase separator); *D* high-flow phase separator; *E* anti-thermoacoustic oscillation volumes; *F* venturi; *G* fill line

strate the ability to control the fluid position during a number of different states of the system: precooling a tank, filling or draining a tank, and during adverse accelerations. To provide these conditions it is therefore necessary to implement an entire system demonstration in orbit. In the design of the space demonstration consideration was given to scaling of the system to larger sizes, hold times of the liquid helium tank before launch, and many other practical limitations. There were also several opportunities that presented themselves in a technology-only mission, namely obtaining engineering data as a primary goal. SHOOT also served as a test bed for intelligent machine control of a complex process in orbit.

The SHOOT payload consisted of two dewars connected by a flexible transfer line. The two dewars, labeled "port" and "starboard", contained removable cryostat inserts ("cryostats") on which the experiments' components, the valves and plumbing were mounted. See Fig. 4.1 for the schematic of one of the SHOOT dewars. The two dewars were identical, but the cryostats contained different LADs (Fig. 4.2) and liquid/vapor detectors to measure the liquid position on orbit. Superfluid helium was pumped between the two dewars using the thermomechanical, or fountain, effect that allowed heat to be directly converted to a pressure differential across a material with fine pores.

During the SHOOT design phase the shuttle Challenger suffered a catastrophic failure during launch. This led to a fundamental rethinking of the way the remaining

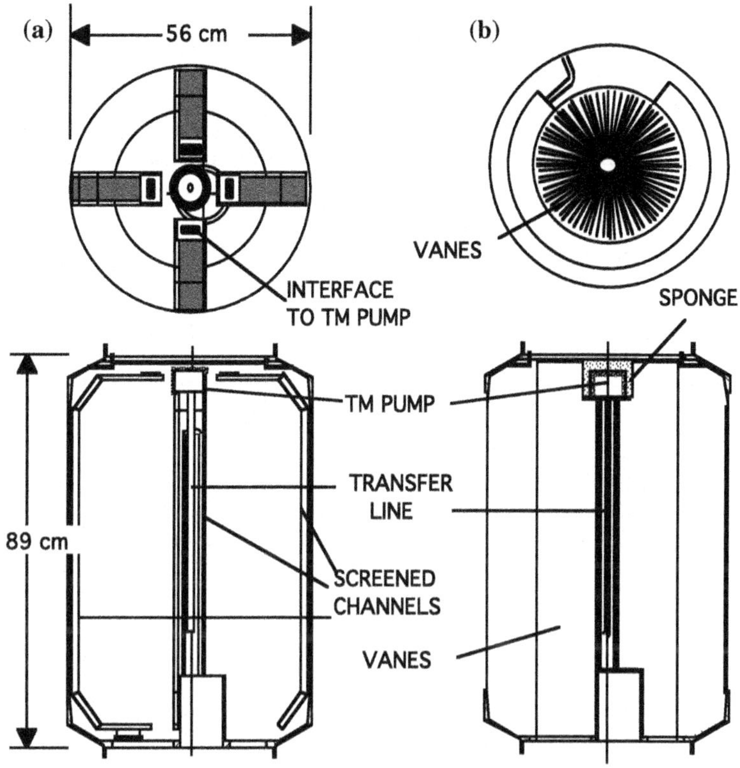


**Fig. 4.2** Arrangement of **a** the screen channels in the starboard helium tank and **b** the vanes and sponge reservoir in the port helium tank. The vanes are fewer in number and extend out to the wall of the helium tank contrary to the schematic illustration

shuttles would be used and to an emphasis on safety. Up until this point the plan was to use SHOOT on two missions: the first to explore the fundamental aspects of helium management in orbit, and the second (SHOOT-II) to explore the human interfaces in a real dewar-to-dewar helium refill. When the shuttle manifest contracted, some of SHOOT-II's goals were folded into SHOOT. A manual, astronaut-Extra Vehicular Activity (EVA) compatible connector that astronauts would exercise during an EVA on SHOOT-II was dropped due to cost and schedule, but an expert system computer program was planned along with an astronaut operated helium transfer from within the shuttle cabin.

In the ensuing years the reality of the cost of space flight and liquid-helium servicing also became apparent—science missions would not be able to afford to resupply on orbit. In most cases the bill was higher than the original science spacecraft cost. The raison d'etre for SHOOT would now be flight-proving the cryogenic components already developed or under development and demonstrating the technology of management of a cryogenic liquid on orbit.

## 4.2 Design Considerations

The SHOOT system was designed for ground testing as well as to function in space. The dewars had to be able to vent while horizontal or vertical. This involved careful positioning of the pumps and liquid/vapor detectors, and adequately sized ground support equipment vacuum pumps. The emergency vent system had to work when surrounded by air or in vacuum, and as far as possible, the operation of the system had to be demonstrated on the ground as well as on orbit.

SHOOT was a short duration space mission (days rather than months or years), which allowed some compromises on heat leaks compared to the extreme thermal isolation required for long life missions. The SHOOT tank was structurally supported from the outer vacuum shell by S-glass composite straps that were sized for infinite fatigue life rather than for the short duration required for testing and launch. This permitted a relatively simple suspension of the two vapor-cooled shields (VCSs) by attaching clamps to the straps. In fact, two VCSs were all that were required rather than the more standard three VCSs, which, again sacrificed thermal performance in favor of simplicity and size of the helium tank. In this configuration with a room temperature outer vacuum shell, the heat leak to the liquid helium was about 180 mW.

### *4.2.1 Structural Requirements*

The dewar structural requirements for inertial, thermal and pressure loads are shown in Table 4.1. The largest inertial load requirement was to survive a very hard landing! (Fig. 4.3).

In addition the dewars were designed as fracture critical hardware, with a fracture control plan and provisions for non-destructive testing. The non-destructive testing included ultrasonic inspection of the raw material per MIL-STD-2154, special dye penetrant inspection per MIL-STD-6866 Type I, Method C or D, and radiography of welds per MIL-STD-453.

The dewars were designed for two launches and landing and a complete qualification test program with a safe-life safety factor of 4.

Bolted hardware was lock-wired to assure that a positive torque was maintained.

**Table 4.1** Inertial loads used in the SHOOT design

| Axis | Preliminary coupled load (Gravity, G) | Updated coupled load (G) | Verification load (G) |
|---|---|---|---|
| X | ±15.7 | ±12.0 | ±5.7 |
| Y | ±4.7 | ±3.8 | ±3.1 |
| Z | ±8.4 | ±6.0 | ±5.5 |

**Fig. 4.3** The space shuttle Endeavor landing a Kennedy space center. The "slapdown" of the nose wheel in a worst case landing is the most severe inertial load for the design

## 4.3 Dewar and Cryostat Details

### *4.3.1 Dewar Fabrication Details*

The SHOOT Payload was mounted to a cross-bay carrier on the space shuttle (see Fig. 4.4).

Each dewar consisted of five main components (Fig. 4.1): the cryostat (which is discussed in the next section), fiberglass support straps, helium tank, inner and outer vapor cooled shields (IVCS/OVCS), and main shell. The dewars were designed to meet the shuttle safety requirements and the cryostats were readily serviceable. The dewar design loads were based on a combination of inertial and thermal loads, the maximum internal pressure of 410 kPa, and the support strap preload. Each load (Table 4.1) was multiplied by an appropriate factor of safety and then applied to a payload model in a NASTRAN finite element structural analysis. In general, highest element loads were extracted for each component so that the component was designed based on those loads.

An uncertainty factor of 1.25 was applied to these loads. The SHOOT dewars and components were designed to the preliminary loads. The SHOOT support structure was designed to the updated loads. The verification load was applied during vibration testing.

Each dewar had three straps on each end that were located 120° apart. The angle of the straps to the tank was chosen to minimize the tank to vacuum shell differential contraction effect on the straps' tension. The straps were fabricated by Structural Composites Industries from Owens-Corning Fiberglass Type S-2 high

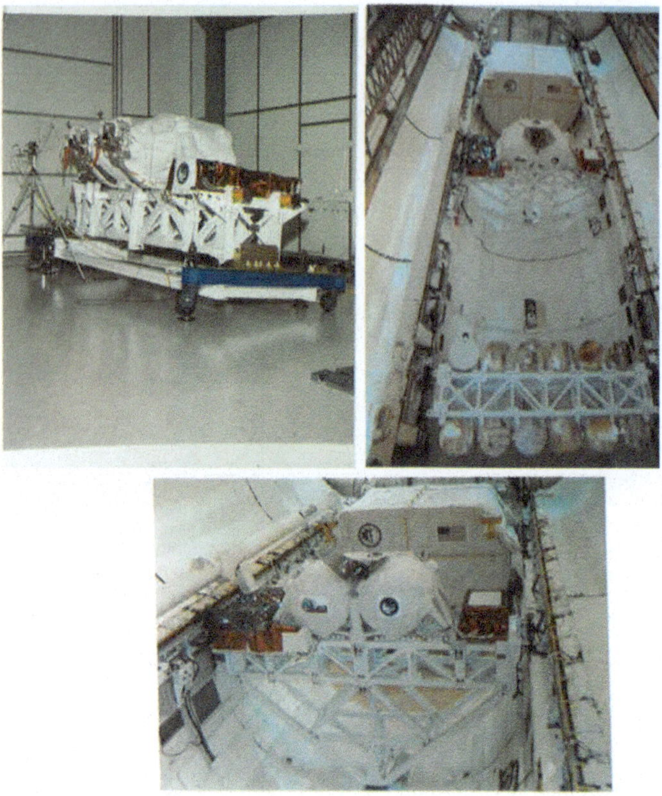

**Fig. 4.4** SHOOT at Kennedy space center. *Upper left* SHOOT on the cross bay bridge. *Upper right* SHOOT mounted in STS-57. *Lower center* SHOOT is mounted near the top of the payload bay just behind the Spacehab module

strength roving, bonded with SCI REZ 081 epoxy resin. The fiberglass was wound around two bobbins with a center-to-center length of 133 mm. The width of each strap was 20 mm and the thickness of each leg was 1.3 mm. The straps could only carry tensile loads, thus a preload of 11,000 N, based on inertial, thermal, and pressure loads, was applied to ensure that the straps were always in tension. The straps were instrumented with two strain gages to accurately measure the applied preload. The straps underwent a maximum of 12,000 cycles during ground test and flight, with a maximum load of 20,000 N. Several straps were tested from the production lot and showed an ultimate strength of greater than 90,000 N and a fatigue life (to the maximum load) of greater than 250,000 cycles. Additional testing showed that there was only a 10 % reduction in ultimate strength after 12,000 cycles [2].

The helium tanks had a 206 L helium capacity. The cold plate of the cryostat was mounted to the helium tank on the forward end with an indium seal and a closeout cover was mounted on the aft end also with an indium seal. The barrel of the helium tank was formed from a rolled cylinder of 3.18 mm thick 2219 T37 aluminum sheet. The two heads were machined from a 15 cm thick plate of 2219 T37 aluminum and welded to each end of the cylinder. The tank then was heat treated to the T87 condition, machined to final dimensions, and qualified by pressurizing to 735 kPa. The stresses that developed from this pressure, envelope the maximum stress due to inertial, thermal, internal pressure, and strap preload which would be seen during flight. Additional units were proof-tested to 575 kPa. During dewar assembly each tank was covered by a five layer multilayer insulation (MLI) blanket, consisting of double aluminized Mylar with Dacron net spacers.

The VCS used the cold, vented gas from the helium tank to intercept heat coming into the dewar. Each SHOOT dewar had two VCSs made from 0.5 mm sheets of 1100 series H14 aluminum. Each VCS was anchored at each strap through an aluminum block that was clamped between the legs of the strap. Epoxy bonding was attempted at first, but peeling of the outer fiberglass layer occurred at the bond edge. The forward anchors allowed flexibility in the axial direction of the tank to prevent buckling of the shields during cooldown. The forward and aft cones of the VCS's were removable to allow access to the cryostat and helium tank. During dewar assembly the inner VCS was covered by a 15 layer MLI blanket, and the outer VCS was covered with a 35 layer MLI blanket.

The main shell contained the cryostat, helium tank, and VCSs and provided attachment points between the dewar and the flight support structure. The barrel of the main shell was a rolled cylinder of 3.18 mm thick 2219 T37 aluminum sheet. The hard points of the main shell, which served as the interface to the flight support structure and the anchor for the support straps were two girth rings. The girth rings were machined from a 15 cm thick plate of 2219 T37 aluminum and welded to each end of the cylinder. The main shell was then heat treated to the T87 condition and machined to final dimensions. The forward and aft heads were spun domes from aluminum 2219 and 6061 respectively. After spinning they were heat treated to the T6 condition and then machined to final dimensions.

### 4.3.2 *Cryostat Details*

The dewar/cryostat system was designed for the easy removal of various components. The majority of these components were in the cryostat. The port and starboard cryostats were nearly identical to allow easy-change out of parts or even the entire assembly. To facilitate the removal of the cryostat, the dewar plumbing and valves were located in a small volume. Thermal attachment of the gas vent lines to the vapor cooled shields was made over a relatively short distance at the top of the dewar. Thermal conduction in the vapor-cooled shield material provided a nearly isothermal environment. For relatively small dewars, as in SHOOT, this scheme led to an acceptable temperature difference along the shields of about 13 K in the outer shield and 2 K in the inner shield.

In the jargon of space helium dewar design, SHOOT had an "instrument dominated" heat load as opposed to a "parasitic dominated" heat load. That is, the heat generated within the helium tank was large enough, on average, that optimization of vapor cooling to reduce parasitic heat did not much increase its lifetime. Coupled with the mission duration of about one week, this allowed the plumbing design to be optimized for the transfer of helium, rather than configured as a long-term storage dewar.

The cryostat plumbing lines consisted of a low-flow vent, high-flow vent, transfer line, and two emergency vents (refer to the schematic in Fig. 4.1). The low-flow vent led from the liquid/gas phase separator within the cryogen tank through a warm valve and out of the dewar. This was the normal vent path out of the dewar and was the primary means of cooling the VCSs. It was made of 12.7 mm diameter stainless steel tubing with 0.5 mm thick walls. The tube was interrupted by thin-walled bellows between the cold plate, IVCS, OVCS, and warm plate to allow for some misalignment and provide structural isolation between dewar components. The total length of the vent was approximately 1.2 m.

A cold valve could be used to bypass the phase separator for chill down of the dewar. The high-flow vent was a much shorter (0.6 m), straighter, and wider (18 mm i.d.) tube which led from the high-flow phase separator out of the warm plate to a large diameter Ball valve. It provided a low impedance vent path at the expense of vapor cooling efficiency for those times when large amounts of heat were being dissipated in the liquid such as during a transfer. The high-flow vent line was only weakly coupled thermally to the VCSs to minimize warming of the gas vented and thus decrease the pressure drop in the line. This, however, also led to an increased heat input to the cryogen tank of about 50 mW in the standby mode. The transfer line left the cryogen tank through two valves, running through the shields to the external coupler and external transfer line. This was similar tubing to the low-flow vent, and was thermally isolated from the shields as was the high-flow vent. The dual emergency vents provided the means of releasing fluid within the cryogen tank in a controlled, safe way following a catastrophic loss of the dewar guard vacuum. See the section on safety for more details.

Over 300 electrical leads traveled from hermetic connectors on the warm plate to the cold region of the cryostat. 90 % of the leads were made of manganin wire, with the remainder of 0.13 mm diameter copper. The leads were heat stationed at each VCS through a connector potted in Stycast 2850. This has proved to be an effective heat sinking method. Approximately 180 leads were fed into the liquid helium tank through 32 pin hermetic connectors. In our experience, these are the largest commercial connectors that can continue to be hermetic at low temperatures.

The main attachment of the cryostat to the cryogen tank was the cold plate at the top of the tank. The cold plate was machined from an explosion welded aluminum-stainless steel sheet [3].

## 4.4 Components

### 4.4.1 Development Notes

Each individual component developed for or used in SHOOT went through a thorough process of testing and qualification before integration into the flight dewar/cryostat system. Several of the components' performance could be visually observed by using glass dewars. We were able to see the phase separation performance of the high-flow phase separator, the low-flow phase separator, the liquid/vapor detectors, and part of the screen channel liquid acquisition system.

### 4.4.2 Phase Separation

The "natural state" of liquid helium in a dewar on the ground is 4.2 K at one atmosphere (4 K at the reduced pressure in Boulder, CO!). By the same token, in orbit where the pressure is practically zero the equilibrium temperature will be less than 2 K (i.e., superfluid). The equilibrium temperature of the liquid will result from a balance between the pressure drop in the vent line plus the phase separator and the boil off rate.

Without a significant acceleration aiding liquid settling, a method to allow boiling to maintain a low temperature and pressure while not exhausting liquid directly must be employed. For superfluid helium, the fountain effect can be used for this purpose. The fountain effect directly converts a temperature gradient into a pressure gradient in the same direction. In the two-fluid model, superfluid $^4$He is composed of two interpenetrating components: a "super" component that flows without viscosity and a "normal" component that has viscosity and behaves as a normal Newtonian fluid. The fountain effect is realized in a porous material whose pores are small enough to significantly restrict the flow of the normal component while allowing the super component to flow freely. A temperature gradient across the porous medium then leads to a gradient in the proportion of normal and super component which gives rise to an osmostic-like pressure gradient. This pressure gradient is in the same direction as the temperature gradient and is about 12 times larger than the vapor pressure gradient in the opposite direction.

The device typically used is called a porous plug which consists of a disk of porous material (usually stainless steel) with small pores ($\sim$a few microns in diameter) through which a temperature and pressure gradient can be developed while venting out one side. Bulk superfluid helium is prevented from passing out of the plug by the fountain effect across the pores. The pores are sized at about 4 microns in effective diameter; large enough to allow some heat conduction out of the tank to maintain a relatively small temperature drop, while small enough to obtain a relatively large fountain effect (up to about 14 kPa). Trade-offs between ultimate temperature in the tank and large-enough fountain pressures usually end

**Fig. 4.5** Picture of two high-flow phase separators. The cylinders are made of porous sintered stainless steel in a reentrant geometry to maximize the flow area and minimize pressure drop across this device

with a porous plug impedance that is on the same order as the impedance of the vent line downstream of the porous plug.

SHOOT used two types of phase separators: a high- and a low-flow phase separator, schematically shown in Fig. 4.1 as D and C, respectively.

The high-flow phase separator had commonly-used porous stainless steel but in a large area and with pores of effective diameter of 8 microns rather than the standard 4 microns (see Fig. 4.5). The 8 microns and large area were necessary to maintain a low temperature in the dewars during superfluid transfer. When transfers were not taking place, SHOOT's temperature was 1.06 K—the coldest thing in space!

As a passenger on the space shuttle, SHOOT had to abide by the access restrictions common to attached payloads. One of those restrictions was that the last servicing needed to be performed about 65 h prior to launch. For superfluid helium payloads on expendable launch vehicles (IRAS, COBE, IRTS, etc.) the last servicing was 12 h before launch. Waiting this extra time on the launch pad and maintaining the liquid below the lambda point would require an on-board vacuum pump [4] or a normal liquid helium guard tank [5]. The former had proven unreliable on two previous shuttle missions and the latter required a large fraction of the space within the vacuum shell. For SHOOT a new approach was implemented: launch with normal liquid helium (He-I, $T > 2.17$ K) and convert to superfluid (He-II, $T < 2.17$ K) on orbit. The added benefit to this approach was that vapor-cooling could be used while on the launch pad even when the liquid was below 4.2 K due to stratification.

The low-flow phase separator was designed to phase separate He-I from its vapor in addition to He-II from its vapor. The flow of He-I out the vent was throttled by closely spaced ($\sim$6 microns), parallel, high conductivity, copper washers. The washers were hollow in the center and helium flowed radially through

**Fig. 4.6** The SHOOT low-flow phase separator. *Square* and *circular* disks alternate in the stack with tiny Kevlar fiber spacers. The bolt at the center top was sued to adjust the clearance between washers

the gaps to the vent line. The gap width in the stack of washers was controlled by an adjustable screw and pressure plate at the top of the device (see Fig. 4.6). The high conductivity copper carried the heat of vaporization of the helium back to the liquid in the center hole. We discovered that even at 2.17 K the Kapitza resistance at this inner boundary was significant. We then crenelated the inner surface to increase the surface area. This decreased the thermal boundary resistance to an acceptable value and this phase separator performed perfectly at pressures up to 110 kPa. As the dewar vented and temperature decreased the flow rate decreased, reaching a minimum at around the lambda point. During on-orbit operation the high-flow phase separator was used in parallel from slightly above the lambda point to continue lowering the temperature. For more details see Refs. [6, 7].

### *4.4.3 Liquid Acquisition*

The LADs were the heart of the SHOOT demonstration. The screen channels (Fig. 4.7) could only be demonstrated on a small scale in one g and the vanes in the port dewar (Fig. 4.8) could not be demonstrated on the ground at all. Their function was to gather the liquid from various places within the tank and feed it to the pump by using surface tension. These are described elsewhere [8].

**Fig. 4.7** One of 4 screen channels in the starboard dewar. The fine mesh screen (*lower half*) ran along the helium tank wall to scavenge as much helium out of the dewar as possible. A row of liquid-vapor detectors can be seen attached to the sheet metal top of the screen channel

**Fig. 4.8** The SHOOT port dewar cryostat insert. The vanes extend to the outer edge of the inside of the tank and feed fluid to the center. At the top of the cryostat is the low flow phase separator housing (*center*) and the high flow phase separator housing (*right*)

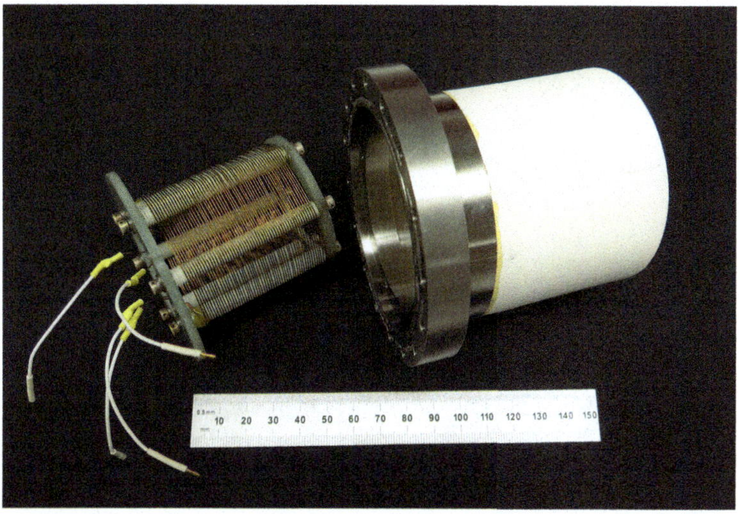


**Fig. 4.9** The SHOOT thermomechanical pump. The item on the left is a manganin wire-wound 50 Ω heater which is inserted into the ceramic cup on the *right*. Together they comprise a pump capable of producing a flow of 1 $m^3$ per hour of superfluid helium

## *4.4.4 Thermomechanical (Fountain Effect) Pumps*

SHOOT used the very simple and reliable fountain effect pumping method. The pumps were simply constructed from a mullite cup having pores with 0.4 micron effective diameter. These pumps were capable of flow of over one $m^3$ per hour and pressures up to 60 kPa [9] (see Fig. 4.9).

Motor driven mechanical pumps were tested with superfluid helium but had difficulty with cavitation at the inlet when a hydrostatic head was not present [10].

### 4.4.4.1 Instrumentation

Knowing roughly where the liquid was during the mission was a key requirement. To solve this problem discreet liquid/vapor detectors were developed. Implementation was very quick (concept to working model in 2 weeks, see Figs. 4.10 and 4.11) and effective [11]. LVDs could be read out by injecting a constant current through a string of Si chips and reading their individual voltages. The response to a change of state was milliseconds and their position accuracy in terms of the liquid-vapor interface was 10s of microns.

The key to understanding the performance of any cryogenic system is thermometry. SHOOT developed a state-of-the-art flight electronics system measured germanium resistance thermometers to 16 bit resolution and accuracy with a 17 Hz excitation. Thermometers were selected for relatively low resistance (<5000 Ω) to

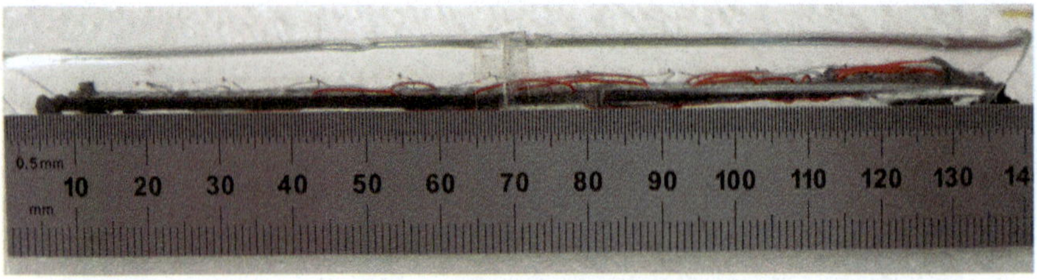


**Fig. 4.10** Test model of the SHOOT liquid/vapor detectors. The *tiny dots* near the top of the picture space 10 mm apart are the 0.25 mm cube Si detectors mounted on a horizontal 0.05 mm diameter stainless steel wire

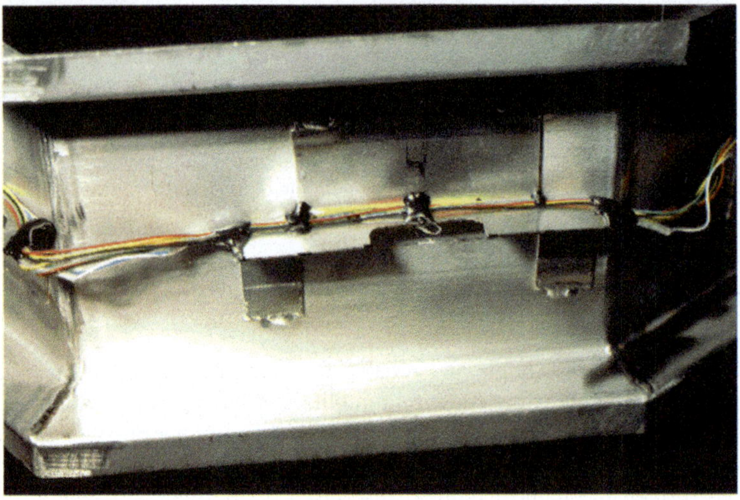

**Fig. 4.11** Close-up picture of one of the liquid-vapor detectors inside the screen channel before the screen is welded in place

mitigate AC parasitic effects. This thermometry enabled high-resolution heat pulse mass gauging as described below.

In low gravity, where the liquid-vapor interface is not well known a priori, the amount of superfluid helium can be simply determined by applying a heat pulse and watching the temperature rise. This technique was used successfully on SHOOT to an accuracy of better than 2 %. It turns out that I learned a bit of forgotten thermodynamics at the same time because most of the 2 % error was due to calculating the fluid amount through enthalpy rather than properly using internal energy!

SHOOT measured the flow rate between the dewars in two ways: by the heating and upstream temperature of the TM pump [9] and by using a venturi flow meter. The venturi was read through variable reluctance pressure transducers and matched the TM pump inferred flow rate to better than 2 %. Integrating the two flow meters provided a check on the heat pulse mass gauging [12].

Strain gauge-based pressure transducers were also used to measure absolute pressure in each cryogen tank. These pressure transducers were bought from Teledyne Taber and used a Wheatstone bridge arrange for the strain gauges which eliminated temperature effects to first order.

### 4.4.5 Cryogenic Stepper-Motor Valves

Stepper-motor driven cryogenic valves for SHOOT were developed and qualified by Ralph Haycock at Utah State University [13]. Two valve sizes were developed, namely a nominal ½ inch valve and nominal ¾ inch valve. The finished valve assembly is shown in Fig. 4.12. The basic valve assembly consists of four major elements: a four-phase stepper motor, a recirculating ball screw assembly, a valve stem load cell, and the valve body. The stepper motor is a four-phase reluctance type motor. The ball bearings in the motors were replaced by dry-film lubricated bearings with increased dimensional tolerance to allow for thermally induced differential contraction. A redundant set of micro-switches at each end position provides positive indication of the valve status and control for the motor. The recirculating ball screw assembly consists of the transmission gears, the recirculating ball screw, and two rigid links that connect the recirculating ball nut to the valve stem through a load cell. The load cell assembly consists of the valve stem and an adjustable load cell consisting of Belleville spring washers stacked in

**Fig. 4.12** The SHOOT stepper-motor-driven cryogenic valve

parallel. The spring washers maintained the stem tip force at cryogenic temperatures despite the differential thermal contraction of the valve elements.

The valves could be operated at any temperature from 300 K down. They required 1 A (1 W at low T) for 30 s generating a total heat input of about 30 J. Two valve sizes were developed, a ¾ inch throughput (based on a Nupro™ ¾ inch valve) and a ½ inch throughput. The larger valve generated a seating force of 202 N. This force caused micro-cracks in the original copper stem when pressed against the stainless steel seat. The copper was replaced with Torlon™ in the flight valves. When sealing against superfluid helium these valves were tested to be leak tight ($<10^{-10}$ Pa $m^3$/s) after more than 100 cycles. The micro-switch position sensors were somewhat unreliable at low temperature. Although they were adjustable at room temperature, they were very sensitive to small changes in position caused by differential contraction.

### 4.4.6 Cryogenic Relief Valves

Each SHOOT dewar had three volumes potentially where trapped liquid could exist between closed valves (Fig. 4.1). Pressure building up in these volumes could result in a vacuum failure. To prevent this, cold relief valves were used. A cold relief valve was developed for SHOOT. A spring held a stainless steel seat against a conical Vespel™ valve head until upstream pressure built in the bellows, moving the stem out. The spring provided a seating force of about 200 N on a seat of 3.8 mm diameter. The cracking pressure was approximately 150 kPa. Tests showed the pressure necessary to achieve full flow was 165 kPa. A number of open-close and thermal cycles were performed, both at room temperature and 4.2 K. In all cases the leak rate at 4.2 K in the closed position was less than $1 \times 10^{-6}$ Pa $m^3$/s. at a pressure difference of 105 kPa, rising to $8 \times 10^{-6}$ Pa $m^3$/s. at 140 kPa.

This valve was not designed to replace burst discs in an emergency venting situation. It had a low throughput designed to handle pressurization of liquid from a nominal heat input of a few watts.

## 4.5 Safety

In 1984, at the beginning of the SHOOT project, we were determined to comply with all of the space shuttle safety requirements to the letter, without asking for waivers. Two-fault tolerance for any failure with major cost or safety impacts was required. To that end, we designed dewars with two dedicated vent lines from the cryogen tank to the outside of the dewar. The dewar cryogen tank was also designed as a "leak-before-burst" vessel. Cold burst disks were located on the cryogen tank and were connected to two warm burst disks on the vacuum shell of each dewar. The outlet of the cold burst disk was hermetically connected to the inlet of the warm burst disk by 19 mm outer diameter (18 mm inner diameter) stainless steel tubes. Thus, there would be no worry about multilayer insulation restricting the flow path through the vacuum space. These dedicated emergency vent lines did add some extra parasitic heat, but since SHOOT's lifetime was to be short on orbit and dominated by heat generated in the transfers, it was an acceptable trade.

The burst disks used for SHOOT were similar to the ones used for COBE [14], but had indium sealed burst diaphragms rather than welded ones. This made it possible for the burst disk activation pressure to be measured and diaphragm punctured before being refurbished for flight.

The burst disks and vent lines were sized to prevent the dewar tank pressure from rising above 413 kPa. An emergency venting analysis was performed to show that the maximum design pressure of the dewar would not be exceeded in the event of a large rupture of the cryogen tank. See Figs. 4.13 and 4.14. Such a large rupture would only be possible if fork-lift tines were rammed through the main vacuum vessel of a dewar.

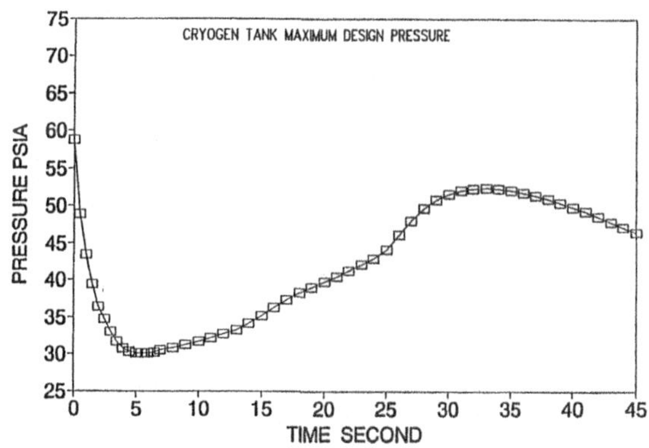


**Fig. 4.13** The predicted blow down pressure in the event of a catastrophic loss of dewar guard vacuum. 60 psia is roughly 4 atmospheres, the pressure at which the cryogen tank burst disks rupture

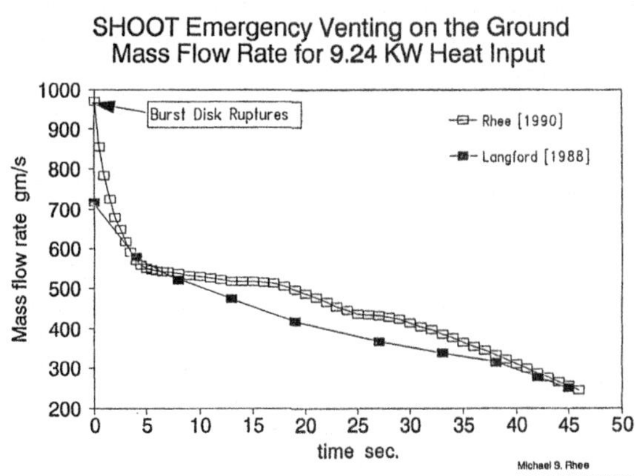


**Fig. 4.14** Two predictions of the predicted helium flow rate versus time out of one emergency vent line after a burst disk rupture

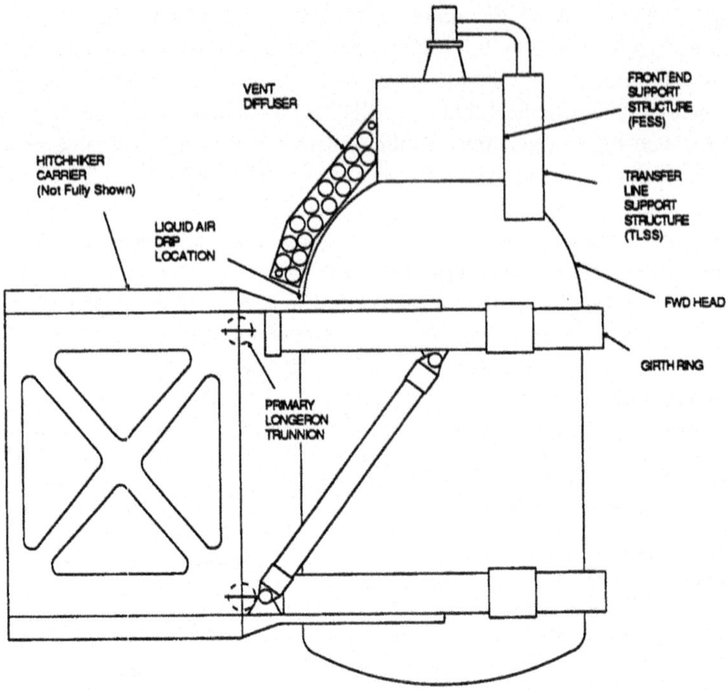


**Fig. 4.15** Sketch of the SHOOT dewar in the orientation in the shuttle bay. The location of any liquid air formation is shown

Each dewar's exhaust helium, for both normal operations and for a sudden loss of vacuum, was channeled into a duct ending in a diffuser that would disperse the cold effluent. See Fig. 4.15. However, in the event of vacuum loss during ground operations, liquid air could form on the outside of the vent duct and diffuser. We tested this possibility by transferring liquid helium into the duct and diffuser at a rate of 400 L per hour for 30 min. Liquid air formation on the outside lower edge of the diffuser was collected and measured. A total of 0.4 L of liquid air was collected over 30 min. To prevent this liquid air from dropping to the floor or into the shuttle bay, a 1 L capacity drip pan was constructed from double aluminized Kapton, a material that is compatible with liquid air. This pan was fastened to the outside of the dewar immediately below the dripping area.

In the final review the Flight Safety panel at Johnson Space Center described our safety design and test effort as "a conspiracy to meet the requirements" with which we happily agreed!

## 4.6 Working with SHOOT on the Ground

The two dewars contained many fill and vent ports which led to an interesting effect that we noticed on the ground. 4.2 K gaseous helium has a density about 1/7th that of liquid so it is relatively heavy. When we had a dewar tipped on its side with one of the vents above the other, the lower vent had He gas outflow, while the upper vent, only a few cm higher was sucking in air!

Near the end of the ground test program one of the external burst disks on one of the dewars developed a small leak. Checking for leaks involved opening a valve that could potentially leak itself. Therefore a leak check cold was not done. After a couple of weeks the dewar was warm as planned and cool down before shipment to the launch site was planned. At this point the burst disks and emergency vent lines were checked. The one emergency vent line could not be evacuated because the cold burst disk on the tank had been ruptured. The rupture was not in the normal direction out of the tank, but in the opposite direction! That is, the leaking external burst disk had allowed enough air into the emergency vent line to freeze a solid ice plug into the line while the dewar was cold. The solid ice plug expanded upon warming, rupturing the burst diaphragm into (instead of out of) the helium tank. This was a very important lesson on the need to leak check thoroughly and as often as possible. This problem cost us 5 weeks of overtime for repairs before shipping to the launch site, but we still arrived in time for our original launch date.

The rubber o-rings that formed a seal into the vacuum space were used in areas of the dewar that were not cold. We did not consider the possibility of diffusion through these o-rings because we carefully ducted vented helium gas away from the dewars. However, given enough time even a few hundred ppm of helium gas in the surrounding atmosphere will contaminate the guard vacuum and lead to a higher heat load. We noticed that over several months without pumping the guard vacuum the heat load increased from 180 to 220 mW.

Another point of discovery was how our low temperature top-off procedure evolved over time. To achieve the maximum fill level on orbit the dewars needed to be topped-off with liquid helium near the superfluid transition. This required pumping down the SHOOT dewars and the supply dewar, well below atmospheric pressure. Similar procedures had been used by several other flight dewars; however SHOOT had differences. SHOOT used two vents on each dewar. We also filled the second SHOOT dewar from the first through the flight transfer line. This meant that we had several flow paths and several vent paths through the top-off. The procedure was rehearsed many times before flight, but each time it required a modification for a new issue: sometimes a vent would produce liquid air, sometimes a vent would be plugged by solid air, and so forth. This continued even on the last servicing on the launch pad. The final servicing concluded successfully, but only by making one final on-the-spot redline to the procedure.

## 4.7 On-Orbit Operations

This case study will not go into many of the details of the SHOOT on orbit results, but a sampling of the measurements made is given here. First, as an experiment, it was a huge advantage for SHOOT to be riding in the space shuttle. Beneficial and adverse accelerations were possible, the communication, power, and data storage were provided, and mission support from the Johnson Space Center and Kennedy Space Center contributed greatly to the success of this Goddard Space Flight Center-led mission.

Second, a lesson about unintended consequences was learned on orbit. SHOOT was launched containing He-I and needed to be pumped down over about 24 h to reach its operating temperature. The plan for this pump down was to use the low-flow phase separator to get close to the superfluid transition, then open the valve to the high-flow phase separator to complete the pump down. This had been tested on the ground to the extent possible, but the high-flow phase separator was difficult to test with a significant hydrostatic head. Thermometers on each phase separator indicated its proper operation: higher temperature on the upstream side of the separator and lower downstream. The port dewar containing the vane system pumped down more quickly than the starboard dewar. Everything appeared normal when the valve upstream of the port dewar's high-flow phase separator was opened, but when the valve on the starboard dewar was opened an alarm rang on the shuttle. The astronauts sleep with the shuttle's nose pointed toward Earth, a gravity-gradient, stable attitude because attitude control thrusters were noisy. The alarm woke the astronauts because the shuttle attitude had swung by 5° from vertical, then swung back to the other side 6°. We learned later that this was caused by a sudden venting of 10s of liters of liquid helium out the vent. Although the flow out the vent was diffuse, the gas impinged on an open shuttle bay that concentrated the momentum of the outflowing gas in the opposite direction. At the time we thought that the high flow phase separator on the starboard tank was broken allowing liquid to escape. What we found out later is that we had such a large pressure gradient and flow rate out of the high-flow phase separator that the upstream liquid was sub-cooled significantly from the liquid in the remainder of the tank [15]. This was the result of a late change to the design that thickened the thin-walled copper heat exchanger around the high-flow phase separator so that the heat flow was restricted. Even superfluid helium can have significant gradients! Discovery of this was hampered by the thermometer wiring being switched on the starboard phase separator, and the different behavior of the port LAD (vanes) which prevented most of the liquid from leaving the port dewar.

After this loss of helium event we had a little over half the amount of liquid we expected to have, but still managed to complete all of the types of transfers planned for the mission. These transfers were: port to starboard high and low rate, starboard to port high and low rate, transfer during adverse accelerations, chill down and transfer into an empty and cold and empty and warm tank, intelligent system controlled transfer, etc.

One of the more interesting side benefits to the expected data were some indications of how slosh in superfluid helium behaved [16]. After a fluid settling acceleration the superfluid motion damped out very quickly with only a couple of large amplitude oscillations.

SHOOT also obtained interesting results on stratification of He-I [17, 18].

## 4.8 Summary

Many components demonstrated for the first time in space, or anywhere for some, came to be used for other projects:

- Stepper-motor valves (XRS, XRS2, SXS and other liquid helium missions)
- Venturi flow meter (to be used on RRM3)
- Mass gauging (ISO and to be used on SXS)
- Fountain pumps (balloon payloads ARCADE and Super-ARCADE)
- Liquid/vapor detectors (shown to also work for other cryogens including liquid nitrogen)
- Cryogenic/ambient burst disks (XRS, XRS2, and SXS)

The SHOOT mission was fully successful in satisfying the original objectives. Along the way several interesting observations were made and lessons learned. The original purpose of helium resupply had disappeared but SHOOT became a demonstration platform for many useful cryogenic components and techniques. SHOOT is still the most extensive cryogenic fluid management experiment conducted in space.

## References

1. P. Kittel, Orbital resupply of liquid helium. J. Spacecraft Rockets, **23**, 391 (1986)
2. K.F. Weintz, M.I. Basci, J.M. Uber, SHOOT Dewar support strap design and performance. Cryogenics **34**, 357–360 (1994)
3. Explosive Fabricators, Incorporated, Louisville, Colorado
4. D. Petrac, U.E. Israelsson, T.S. Luchik, The lambda point experiment: helium cryostat, cryo-servicing, functions, and performance. Adv. Cry. Eng. **39**, 137–144 (1994)
5. R.T. Parmley, J. Goodman, M. Regelbrugge, S. Yuan, Gravity Probe B dewar/probe concept. Proc. of the SPIE **619**, 126 (1986)
6. P.J. Shirron, J.L. Zahniser, M. J. DiPirro (eds.), A liquid/gas phase separator for He-I and He-II. Adv. Cryog. Eng. **37**, 105 (1992)
7. J.G. Tuttle, M.J. DiPirro, P.J. Shirron, Liquid/gas phase separators for the superfluid helium On-orbit transfer (SHOOT) project. Adv. Cryog. Eng. **39**, 121 (1994)
8. P.J. Shirron, M.J. DiPirro, J.G. Tuttle, Flight performance of the SHOOT liquid acquisition devices. Cryogenics **34**, 361 (1994)

9. M.J. DiPirro, E.R. Quinn and R.F. Boyle, Tests of a nearly ideal, high rate thermomechanical pump, in *Proceedings of 12th International Cryogenic Engineering Conference* (Butterworths, London, 1988), p. 646
10. P.R. Ludtke, D.R. Daney, Cavitation characteristics of a small centrifugal pump in He I and He II. Cryogenics **28**, 96 (1988)
11. P.J. Shirron, J.G. Tuttle, M.J. DiPirro, Performance of discrete liquid helium/vapor and He-I/He-II discriminators. Adv. Cryog. Eng. **39**, 1105 (1994)
12. M.J. DiPirro, P.J. Shirron and J.G. Tuttle, Mass gauging and thermometry on the superfluid helium on-orbit transfer flight demonstration. Adv. Cryo. Eng. **39**, 129 (1994)
13. These are now a commercial product of ATK/Mission Research
14. The COBE burst disks were built by Ametek Straza which became Katema. Currently a version is marketed by Hydrodyne
15. M.J. DiPirro, P.J. Shirron, J.G. Tuttle, Superfluid helium transfer in space. Cryogenics **34** (ICEC Supplement), 267 (1994)
16. M.J. DiPirro, P.J. Shirron, J.G. Tuttle, On orbit superfluid transfer: preliminary results of the SHOOT Flight Demonstration. Cryogenics **34**, 349 (1994)
17. P.J. Shirron, M.J. DiPirro, Low gravity thermal stratification of liquid helium on SHOOT. Cryogenics **32**, 159 (1992)
18. J.G. Tuttle, M.J. DiPirro, P.J. Shirron, Thermal stratification of liquid helium in the SHOOT dewars. Cryogenics **34**, 369 (1994)

# Chapter 5
# TESLA & ILC Cryomodules

**T.J. Peterson and J.G. Weisend II**

**Abstract** The Superconducting Radio Frequency (SRF) cryomodules for the TESLA and International Linear Collider (ILC) projects have undergone significant design evolution and prototype testing. Given the large number of cryomodules required for these projects, key requirements include low cost and low heat leak. This chapter provides details of the of the ILC cryomodule design and traces its evolution. Topics include: requirements, He II cooling of the SRF cavities, mechanical (including alignment) and thermal design, fluid flow, safety and vacuum systems. The results of early prototype testing and their impact on the design are also described. Changes in the design to enable its use in the LCLS II accelerator project are also described.

## 5.1 Introduction

The Deutsches Electronen Synchrotron (DESY) in Hamburg, Germany, led the early electron-positron linear collider collaboration which developed the concept called TESLA (Tera-Electronvolt Superconducting Linear Accelerator) [1] in the late 1980s and early 1990s. The TESLA collaboration desired cryomodules with a lower cost per unit length than had been developed up to that time. Among the cost-reducing features would be to connect cryomodules in continuous, long strings similar to cryostats for long strings of superconducting magnets (see for example Chaps. 2 and 3). The elimination of the external cryogenic transfer line by placing all cryogenic supply and return services in the cryomodule could also reduce costs, not only directly for the cryogenic components, but also by reducing tunnel space required.

T.J. Peterson (✉)
Cryomodule Engineering, SRF Development Department, Technical Division,
Fermi National Accelerator Laboratory, P.O. Box 500, Batavia, IL 60510, USA
e-mail: tommy@fnal.gov

J.G. Weisend II
European Spallation Source ERIC, Lund University, P.O. Box 176, 22100 Lund, Sweden
e-mail: john.weisend@esss.se



J.G. Weisend II (ed.), *Cryostat Design*, International Cryogenics Monograph Series,
DOI 10.1007/978-3-319-31150-0_5

This approach, driven mainly by the number of cryomodules required, as many as 2000 in some accelerator concepts [1–3], is in contrast to the segmented cryomodules described in Chap. 6. Segmenting the cryomodules and separating them by warm sections containing magnets, vacuum systems and beam instrumentation tends to be the preferred solution in systems in which fewer than 100 cryomodules are required. The warm sections allow easier access to various accelerator systems. However, the numerous warm—cold transitions along the length of the accelerator result in higher cost and heat leak than can be tolerated in systems the size of TESLA.

Prototype TESLA cryomodules were incorporated into a test linac called the TESLA Test Facility (TTF) at DESY [4]. TTF was modified and expanded into a light source facility called the Free-electron LASer in Hamburg (FLASH). TTF and FLASH demonstrated successful operation of TESLA-style cryomodules in a linac with beam.

Various projects subsequently adopted the TESLA cryomodule design concept with some modifications depending on specific requirements and innovations. These projects include the International Linear Collider (ILC) [2, 3], X-ray Free Electron Laser (XFEL) at DESY [5], and the Linac Coherent Light Source II (LCLS-II) [6] at SLAC. LCLS-II includes more extensive modifications than the other projects described here due to LCLS-II being a continuous wave (CW) linac with higher dynamic heat loads than the other pulsed linac designs. This chapter describes the TESLA-style cryomodule design in general, with examples from TESLA prototype cryomodules and the various subsequent projects. Figure 5.1 illustrates one 12 m long cryomodule mechanical assembly.

## 5.2 Definitions

See Table 5.1.

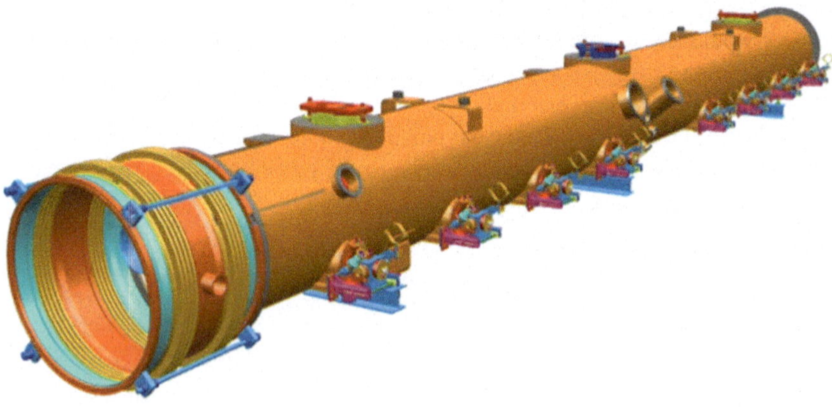

**Fig. 5.1** ILC cryomodule assembly showing vacuum vessel, eight RF power input couplers, three internal supports on top, and open ports for instrumentation flanges and for power leads providing current to a centrally located superconducting magnet package

**Table 5.1** Definitions

| Term | Definition |
|---|---|
| SRF | Superconducting radio frequency |
| CW | Continuous wave operating mode |
| HOM | Higher order mode |
| TESLA | Tera-electronvolt superconducting linear accelerator (TESLA)—the predecessor to the ILC concept, developed collaboratively by DESY, Saclay, INFN, Fermilab, Jefferson Lab, Cornell, and other labs and universities under the leadership of DESY. The TESLA collaboration developed cryomodules with RF cavities in close-fitting helium vessels suspended from a large helium return pipe, cryogenic piping all within the cryostat. Hence, cryomodules having that configuration are referred to here as TESLA-style |
| ILC | International linear collider |
| XFEL | X-ray free electron laser, an SRF linac at DESY in Hamburg, Germany, which includes TESLA-style cryomodules |
| LCLS-II | Linac coherent light source II, which includes a 4 GeV superconducting linac at SLAC utilizing TESLA/ILC-style cryomodules modified for CW operation in the existing SLAC tunnel |
| 2 K | Nominally 2 K temperature level, typically 1.8–2.1 K |
| 5 K | Nominally 5 K temperature level, which includes temperatures approximately from 4.5 to 7.5 K |
| 70 K | Nominally 70 K temperature level, which ranges from approximately 35–85 K. (Note that cryogenic circuits, thermal shields, and thermal intercepts will be $\sim$80 K when LN2 is the coolant. Lower temperatures are provided by gaseous helium.) |
| MAWP | Maximum Allowable Working Pressure, a term that is used to define the safe pressure rating of a component or test system |
| Cold mass | Those portions of the cryomodule within the vacuum vessel which are cooler than room-temperature—RF cavities, piping, thermal shield, etc. |
| HGRP | Helium gas return pipe, the 300 mm diameter helium pipe which also serves as the structural backbone of the cold mass in TESLA-style cryomodules |
| MLI | Multilayer insulation, or "superinsulation", aluminized mylar wrapped in layers alternately with a conductively insulating material in the insulating vacuum space to block thermal radiation |
| Q0 | Cavity quality factor, of interest for cryomodule design since power dissipated by the RF cavity to the nominally 2 K temperature level is inversely proportional to Q0 |

## 5.3 Functional Requirements Summary

The TESLA and ILC main linac cryomodules contain superconducting cavities together with cryogenic distribution lines, thermal shielding, magnets, and instrumentation. Most fundamentally, the cryomodule provides insulating vacuum, cooling, RF power, and RF cavity support and alignment for the proper functioning of the SRF cavities in the linac.

Strings without breaks between individual cryomodules include up to 12 cryomodules. In addition to niobium superconducting RF cavities, cryomodules may include a combined focusing and steering magnet, a beam position monitor, and gate valves on the beamline for vacuum isolation during assembly. Beam line higher-order-mode (HOM) absorbers are located in the cold beamline interconnects between cryomodules. The cavity tuners, HOM couplers, and fundamental power couplers attach to each cavity. Cryogenic transfer lines are required only where bypasses for warm sections of the linac must be provided.

TESLA-style cryomodules have been fabricated with four 3.9 GHz RF cavities as well as with eight 1.3 GHz RF cavities. The ILC concept included some eight-cavity cryomodules with one magnet package (the configuration illustrated in Fig. 5.1) and some nine-cavity cryomodules without a magnet. The final TESLA cryomodule design, as described in the TESLA TDR [1], contained 12 RF cavities in a further attempt to improve the packing factor of RF cavities along the linac.

A mechanical tuner with motor-driven and piezo-driven components provides active control of the RF cavity resonant frequency. RF cavities are independently powered through a fundamental power coupler on each cavity connected via air-filled waveguide to an RF source (klystron or solid-state amplifier).

Cryogenic circuits provide 2 K liquid helium to the cavities with a valve for liquid supply in each string of cryomodules or, in the case of LCLS-II, via a valve in each cryomodule. Managing a liquid helium bath of up to 100 m or so in length is one factor in determining string length. The cryostat includes a nominally 5 K thermal intercept circuit, which may include a thermal radiation shield, and a nominally 70 K thermal radiation shield and thermal intercept circuit. Dynamic RF heating of the cavities at 2 K is a major contributor to overall cryogenic refrigeration required.

The Table 5.2 provides a summary of the key functional requirements for a TESLA/ILC-style cryomodule that may drive the design. These are topics, which require particular attention during design.

## 5.4 Cryomodule Mechanical Design

### *5.4.1 Cryomodule Major Components and Features*

Cryomodules consist of various complex subassemblies, which we describe in this section, starting from the inside with the niobium RF cavity and working out to the

**Table 5.2** Key functional requirements for the cryomodules

| Key requirement | Description |
|---|---|
| Series configuration | Cryomodules with insulating vacuum open at each end to the next, so connected insulating vacuums, and with cold beam pipe through the interconnect. This configuration implies considerations for the in situ connection of not only the accelerator beamline but also cryogenic pipes and insulating vacuum |
| No external parallel transfer line | Cryomodules include all cryogenic piping within the cryostat, as opposed to having a parallel external cryogenic transfer line to supply cooling to cryomodules |
| Microphonics | As for any SRF cavity support structure, minimize cavity vibration and coupling of external sources to cavities. This is addressed by means of providing a stiff support system and stiffening of elements such as the thermal shield |
| Alignment | Provide good cavity, quadrupole, and BPM alignment (typically <0.5 mm RMS) |
| Vibration | Minimize quadrupole magnet vibration for gradient stability of typically 0.01 % RMS or better |
| Seismic design | In many locations (especially Japan and California), one must design in accordance with local seismic requirements |
| Thermal efficiency | Intercept significant heat loads at intermediate temperatures above 2.0 K to the extent possible. Intercepts include not only thermal radiation shield or shields and support structures, but may also include intercepts for large cables, magnet current leads, and RF components such as HOM couplers and absorbers |
| Pressure safety | Cryomodules and components must comply with the requirements of the institution and/or locality in which they will operate. For accelerator laboratories in the US, this means the Department of Energy regulation 10CFR851—a level of safety equivalent to pressure code safety level. Designers must protect the helium and vacuum spaces including the RF cavity from exceeding MAWP |
| Magnetic shielding | Provide excellent magnetic shielding and very low residual field (for example, ≤5 mG) at the niobium cavities to preserve high Q0 for less power lost as heat to 2 K |
| Thermal performance | For RF cavities operating in CW mode, one may have to allow removal of over 20 W per cavity and/or 150 W at 2 K per cryomodule |
| Cool-down | Provide, to the extent possible given the cryomodule string configuration, cool-down conditions which retain high cavity Q0. Recent recognition of the role of thermal gradient on the niobium cavity to "sweep out" magnetic flux as the Nb passes through the superconducting transition during cool-down may create new design goals for management of cool-down, such as the addition of a cool-down valve in each cryomodule where retaining high Q0 is particularly important |

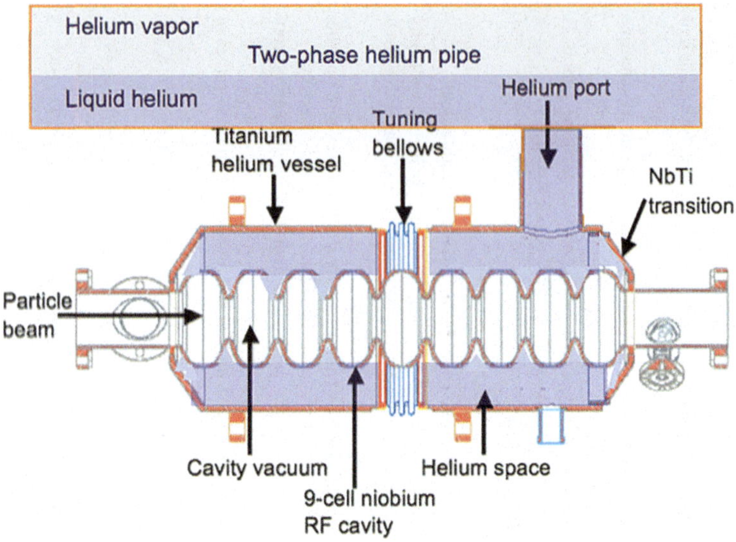


**Fig. 5.2** Helium vessel concept (not to scale) for an ILC 9-cell cavity

vacuum vessel. Figure 5.2 shows a sketch (not to scale) of the titanium helium vessel assembly surrounding a 1.3 GHz niobium cavity consisting of multiple elliptical cells. Liquid helium completely fills the helium vessel, surrounding the RF cavity with a liquid level into the 2-phase pipe.

Heat flows via helium II heat transport through the liquid helium to the liquid surface in the 2-phase pipe where the helium evaporates. Thus, cooling below the lambda point involves no bubbling, an advantage in reducing microphonics disturbance of the RF cavity.

Figures 5.3 and 5.4 illustrate the dressed cavity and a portion of the RF cavity string. Each of the cavities is encased in a titanium helium vessel, supported from

**Fig. 5.3** Dressed cavity. The titanium helium vessel is shown as transparent for a view of the 9-cell niobium RF cavity inside. This image includes an original design version of the Blade Tuner and shows the 2-phase pipe connection on top of the helium vessel. The connection from the helium vessel to the 2-phase pipe is called the nozzle or "chimney"

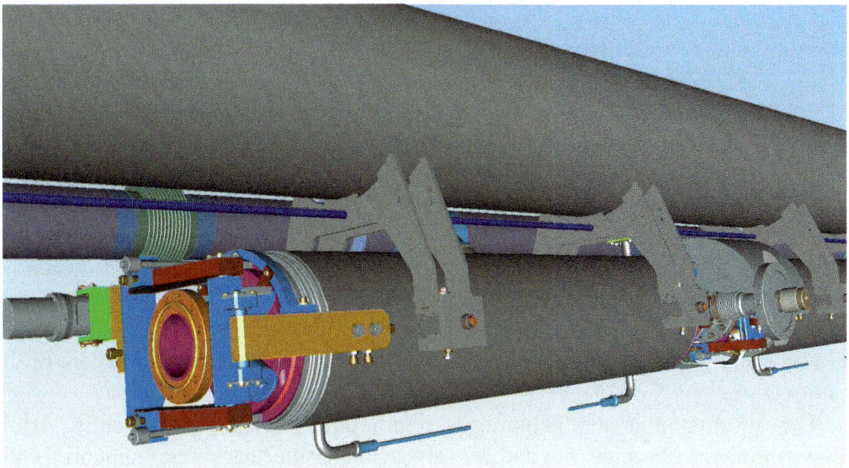

**Fig. 5.4** Dressed cavity as part of the cavity string within a cryomodule for LCLS-II. Other features shown include the RF power coupler for one cavity, an end lever tuner for an LCLS-II dressed cavity, support arms for the dressed cavities, and the invar rod just below the HGRP which provides the axial position anchor for the dressed cavities

the HGRP by means of stainless steel brackets connected to four titanium pads on the helium vessel itself. Each bracket is equipped with a longitudinal sliding mechanism and adjusting screws and pushers for alignment. A mechanical, coaxial (blade) tuner and a piezo-electric tuner are mounted to the vessel.

Dressed cavities are connected to one another via the beam tube connections and the 2-phase pipe connections. An ILC cryomodule includes a string of eight or nine such cavities connected in series. In the case of eight cavities, the ninth element in the string is a corrector magnet package, also cooled at nominally 2 K. With the cavities and magnet package each being about 1.3 m in length, the entire cryomodule assembly is about 12 m long.

The inter-cavity spacing—which accommodates RF- and HOM-couplers and a flanged interconnecting bellows—amounts to approximately 300 mm, depending on mechanical and RF details. For example, desire to operate beam in either direction may dictate even-wavelength spacing. Conversely, desire to disrupt the flow of "dark current" (electrons from field emission) may dictate non-integer wavelength spacing. Manually operated valves required by the clean-room assembly terminate the beam pipe at both module ends. The valves are fitted with simple RF shields (Fig. 5.5).

Magnetic shielding surrounds the dressed cavity in order to limit imposed magnetic field on the niobium superconductor. One must watch also for internal sources of magnetic fields. No component of the cryomodule should impose more than a specified low magnetic field (such as 5 milligauss, the specification for LCLS-II) on the niobium cavity during cool-down through the 9.2 K transition temperature, since trapped flux then reduces the efficiency of the RF cavity,

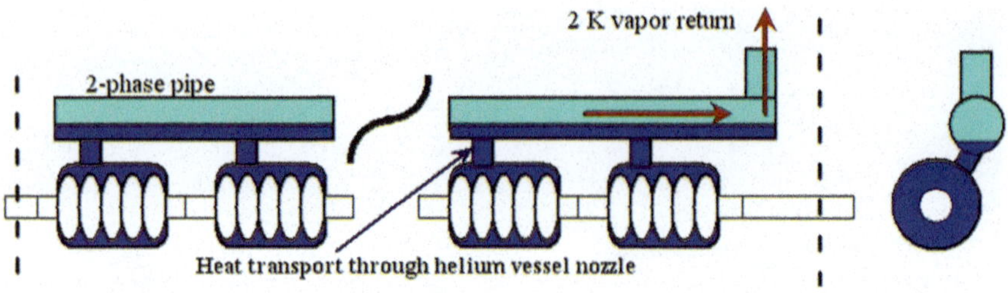


**Fig. 5.5** Cooling scheme for an ILC cavity string

resulting in higher heat loads at 2 K and/or limiting peak operating gradient of the cavity.

The 300 mm diameter helium gas return pipe (HGRP), prominently visible above the cavities in Figs. 5.6 and 5.7 serves both as the "backbone" support for the RF cavity string in each cryomodule and also as the low pressure, 2 K, saturated vapor return line to the cryogenic distribution box. The HGRP is support by three epoxy-fiberglass posts [7–9]. To accommodate the HGRP thermal contraction when cold relative to the vacuum vessel, the two side post brackets can slide over the top flanges while the central post bracket is locked in position.

The support posts (see Figs. 5.7 and 5.8) consist of a fiberglass tube terminated by two shrink-fit stainless steel flanges. Two additional shrink-fit aluminum flanges are provided to allow intermediate thermal intercept connections to the 5–8 K and

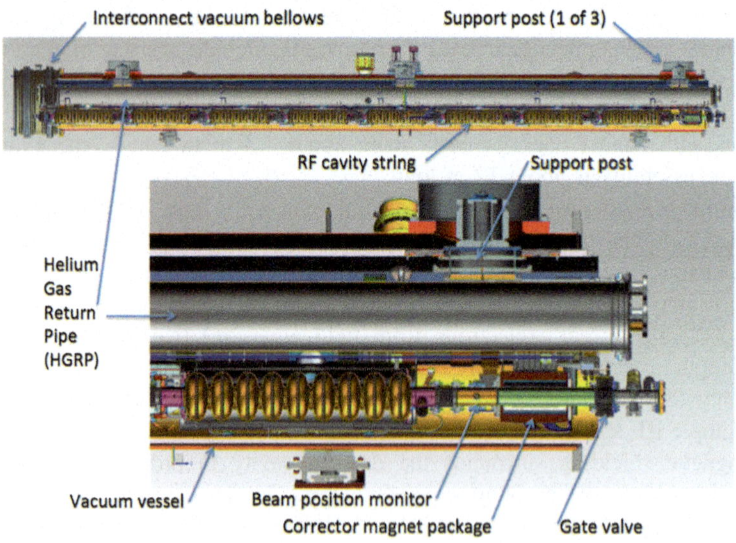


**Fig. 5.6** Some major features of a TESLA/ILC-style cryomodule. This particular image is of an LCLS-II 1.3 GHz cryomodule assembly, but the features shown are typical of TESLA/ILC-style cryomodules (*credit* for figure—Fermilab)

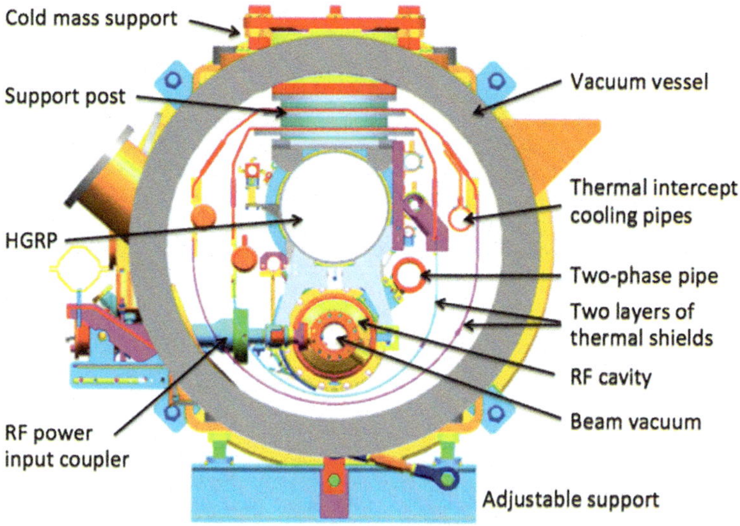


**Fig. 5.7** Cross-section of an ILC cryomodule showing major features which are typical of a TESLA-style cryomodule

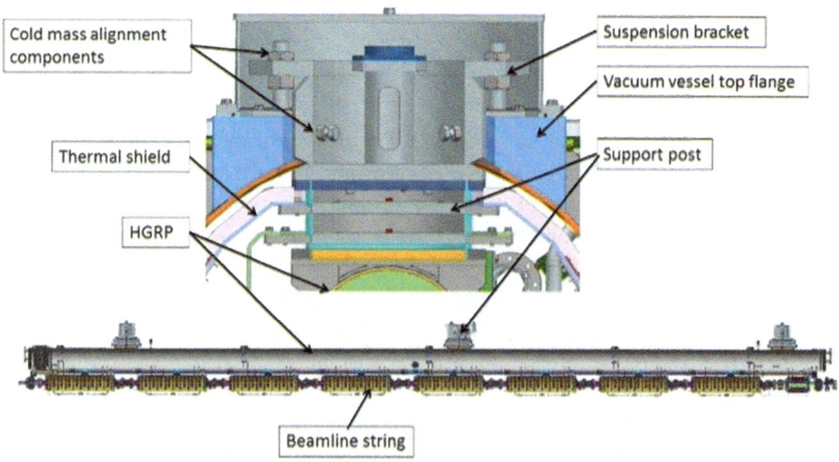


**Fig. 5.8** Cold mass support and alignment system (*credit* for figure—Fermilab)

nominally 70 K temperature levels; the exact location of these flanges has been optimized to minimize the heat leakage.

RF cavities are supported from the HGRP by means of four lugs on the helium vessel. Stainless steel arms welded to the HGRP extend down to C-shaped holders containing needle bearings, shown in Fig. 5.9. The cavities are anchored in position

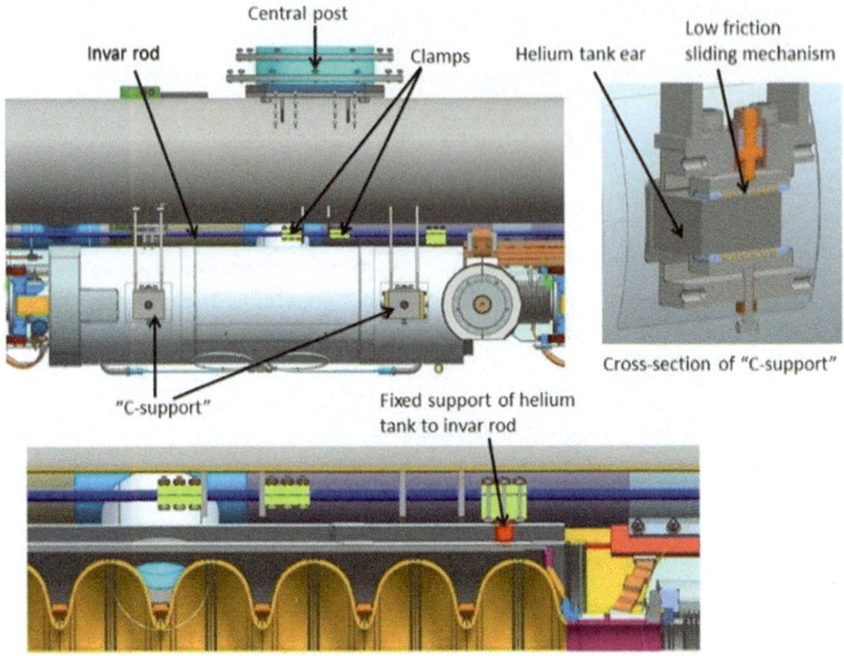


**Fig. 5.9** Support system of cavity helium tank to the HGRP and invar rod (*credit* for figure—Fermilab)

longitudinally via a clamp to an invar rod. The needle bearings allow shrinkage of the stainless HGRP relative to the RF cavities and magnet package.

Cavity positions relative to fiducials on the vacuum vessel are set during assembly with no requirement for later internal adjustment of cavity position within the cryomodule after assembly. Alignment needs to be maintained with cryomodule transport, thermal cycling, and pressure cycling. (For example, return to position within 0.5 mm RMS tolerance for LCLS-II.) Final alignment requires positioning of the vacuum vessel assembly with reference to the external fiducials, which were in turn referenced to the cavity string. Table 5.3 summarizes approximate allocation of alignment tolerances, illustrating that for each source of misalignment, due to their additive nature, tolerances must be tighter than the overall requirement.

Linac beam operational requirements determine cryomodule lattice dimensions and intracavity spacing. Since the linac is operated cold, but the cryomodule is assembled warm, careful analysis of thermal contractions is required for the design of warm assembly dimensions. Key dimensions include main coupler cavity-to-cavity distance and cryomodule slot length.

RF power travels via waveguide and connections to the cryomodule input couplers. Various other electrical connectors are located on round cover plates mounted on the vacuum shell, the connector plates sealed with O-ring seals. A pair

**Table 5.3** Typical alignment tolerances for RF cavities within a cryomodule

| | Subassembly | Tolerances (RMS) (mm) | Total envelope |
|---|---|---|---|
| Cryomodule assembly | Cavity and helium vessel | ±0.1 | Positioning of the cavity with respect to external reference ±0.5 mm |
| | Supporting system | ±0.2 | |
| | Vacuum vessel construction | ±0.2 | |
| | Action | | |
| Transport, testing, and operation | Transport and handling (±0.5 g in any direction) | ±0.2 | Reproducibility and stability of the cavity position with respect to external reference ±0.5 mm |
| | Vacuum pumping | ±0.3 | |
| | Cool-down | | |
| | RF tests | | |
| | Warm-up | | |
| | Thermal cycles | | |

of instrumentation flanges is also associated with each input coupler to reduce wire lengths and risks associated with long runs of wires inside the cryomodule.

Figures 5.10 and 5.11 illustrate the region in the cryomodule around the superconducting magnet and one example (for a Type 3+ cryomodule) of a magnet package. Like for the RF cavities, the 300 mm diameter HGRP supports the superconducting magnet package, consisting of a quadrupole and two correction dipoles. Also like for the RF cavities, the invar (see Chap. 1) rod maintains the magnet package axial position, while the HGRP supports the magnet mass and provides lateral and vertical positioning. A beam position monitor is pinned to the magnet package for precise relative alignment of the two. Current leads shown in Figs. 5.10 and 5.11 are conductively cooled, based on the design developed by CERN for the LHC corrector magnet current leads [10]. Conductive cooling has the advantage of not requiring helium gas flow, which in these cryomodules one could not provide from the 2-Kelvin magnet coolant due to the low (nominally 30 mbar) pressure. Relatively low currents (100 A or less) enable the use of conductive cooling without too much heat input to the low temperature levels of the cryomodule.

The cryostat for TESLA, ILC, and XFEL includes two aluminum radiation shields operating in the temperature range of 5–8 K and 40–80 K respectively. For LCLS-II, the nominally 5 K thermal shield is eliminated although the helium circuit is retained to provide 5 K thermal intercepts for RF power input couplers, magnet current leads, support posts, and some RF cables. Thermal radiation shield bridges are provided through each cryomodule interconnect.

Blankets of multilayer insulation (MLI) are placed on the outside of the thermal shields and other cold surfaces. At the nominally 70 K level, MLI blankets consist of typically 30 layers. At the 2 and 5 K levels, MLI blankets consist of 5–10 layers

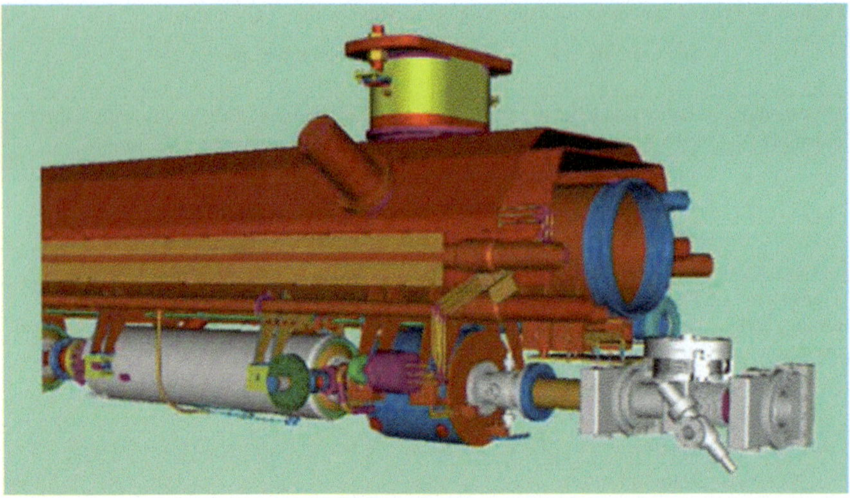

**Fig. 5.10** A "Type 3+" cryomodule, very similar to those for XFEL, showing from the left below the 300 mm pipe: the last dressed cavity in the string, the magnet package, BPM, gate valve, HOM absorber, and another vacuum gate valve (*credit* for image: XFEL TDR [5])

**Fig. 5.11** A Type 3+ cryomodule magnet package, showing current leads, helium vessel containing the magnet, beam position monitor attached to magnet end, and 2-phase pipe section associated with the magnet (*credit* for image: XFEL TDR [5])

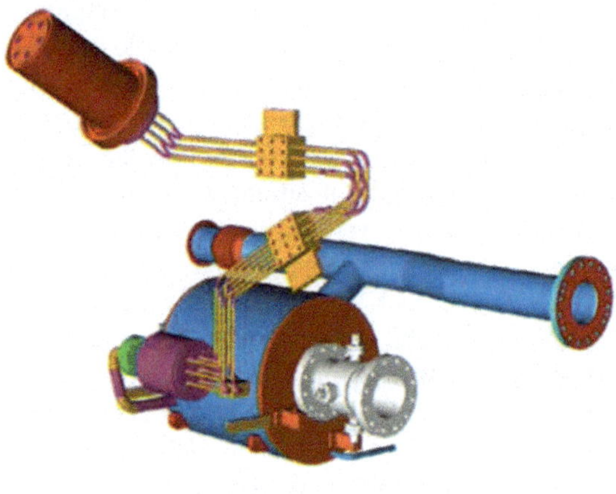

while the nominally 70 K blanket contains 30 layers. Cavity and quadrupole helium vessels, gas return pipe and 5–8 K pipes are wrapped with 5 to 10 layers of MLI to reduce heat transfer in the event of a vacuum failure.

Thermal shields are constructed from a stiff upper part (divided into two halves), and multiple lower sections (according to the number of the cold active components, e.g. cavities, magnets). Intermediate flanges on the fiberglass posts support the thermal shields from the top. Thermal shields are screwed to the center post but can axially slide on the other two posts, to which they are still thermally connected, allowing for thermal contraction relative to the vacuum vessel. Slots cut into the

**Table 5.4** Approximate typical cryomodule mass (kg)

| Cryomodule component | Mass (kg) |
|---|---|
| Cold mass (cavity string, thermal shield, HGRP, other cold pipes) | 3300 |
| Vacuum vessel | 3300 |
| End flanges, reinforcing rings | 300 |
| HGRP supports (posts, brackets, support covers, 3 sets) | 900 |
| Other items (vacuum bellows, external support base, power couplers, instrumentation, etc.) | 800 |
| Total | 8600 |

thermal shields near cooling pipe attachments reduce bowing due to differential thermal contraction during cool-down and warm-up.

### 5.4.2 *Cryomodule Weight*

Table 5.4, lists cryomodule component and total weights, in particular for the LCLS-II cryomodule. ILC would be very similar. The bottom line is 8600 kg for the cryomodule assembly.

### 5.4.3 *Major Interfaces*

Major interfaces from the cryomodule to other linac components include a wave guide connection to the RF power input coupler, instrumentation connectors, magnet power lead connections, the anchors from the adjustable supports to the floor, pump connections to the insulating vacuum pump-out ports, and may also include guard helium vacuum connections for helium-shrouded subatmospheric connections.

## 5.5 Cryomodule Vacuum Design and Vacuum Vessel

The cryostat outer vacuum vessel is constructed from carbon steel and, for these TESLA-style cryomodules, has a standard diameter of 38 inches. Adjacent vacuum vessels are connected to each other by means of a cylindrical sleeve with a bellows, sealed to each cryomodule end by an O-ring vacuum seal. In the event of accidental spills of liquid helium from the cavity vessels, a relief valve on the sleeve together with venting holes on the shields prevent excessive pressure build-up in the vacuum vessel.

Wires and cables of each module pass through the vacuum shell via metallic sealed flanges with vacuum tight connectors. The insulating vacuum system is pumped during normal operation by permanent pump stations located at appropriate intervals. Additional pumping ports are available for movable pump carts, which are used for initial pump down, and in the event of a helium leak. Due to isolation of the cavity vacuum at the RF power coupler by means of a ceramic window, the RF power coupler typically needs an additional vacuum system on its room temperature side; this is provided by a common pump line for all couplers in a module, which is equipped with an ion getter and a titanium sublimation pump.

A new feature for the LCLS-II cryomodules in particular is access ports for the tuners, serving to provide access to the piezo tuners within the cryomodules. This feature eliminates the need to pull the entire cold mass string out of the vacuum vessel in order to access tuner components.

The insulating vacuum is protected from over pressurization by means of a spring-loaded lift plate. A single worst case piping rupture internal to the insulating vacuum must be analyzed to determine lift plate size. Provisions are provided to allow free passage of the helium out past thermal shield and MLI to the lift plate, which is typically 200–300 mm in diameter.

The cavity (beam) vacuum must be kept very clean and particle free. The vacuum gate valves shown in Fig. 5.10 provide beam vacuum isolation during cryostat assembly and interconnect assembly. In the linac, beam vacuum pumping and readouts are provided at warm beamline sections. Warm beamline vacuum components adjacent to the cryomodule cold beam vacuum must be clean and particle free to avoid contamination of the cavity vacuum. Pumping may be provided by titanium sublimation pumps, with these at the ends of cryomodule strings providing "guard" pumping to reduce cryopumping of the warm beamline into the cold RF cavity vacuum. Ion pumps may provide vacuum readout

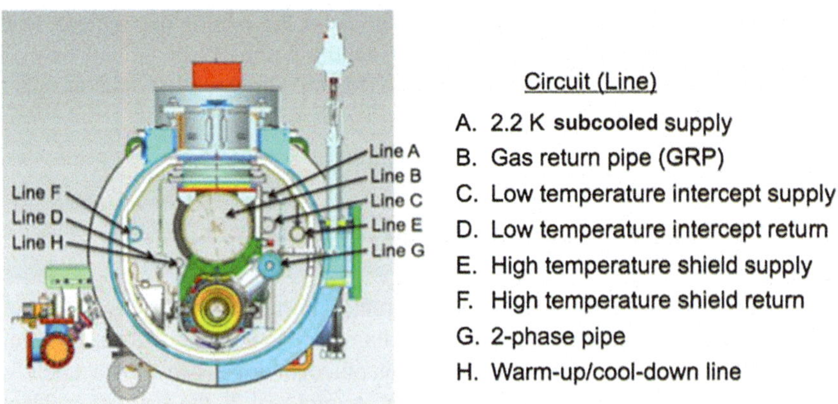


**Fig. 5.12** LCLS-II cryomodule assembly cross-section

## 5.6 Cryomodule Thermal Design and Helium Flow Design

### *5.6.1 Major Thermal Design Features*

The RF cavities are maintained at nominally 2 K by means of a stagnant bath of saturated liquid helium. Temperatures of 1.8–2.1 K are feasible in a large system, and the RF cavity helium vessel and piping design accommodate any temperature within this range. The 2 K helium is supplied by a pressurized, nominally 2 K line, to the cryomodule or string liquid level control valves.

A nominally 5 K cooling circuit will typically incorporate pressures above the helium critical pressure (2.27 bar) so as to avoid 2-phase flow. Although no phase change takes place as the supercritical helium warms, in the temperature range of 5–8 K as much as 30 J/g may be absorbed, comparable to that of a phase transition. A 5 K circuit provides low temperature thermal intercept for the support posts, magnet current leads, RF power coupler, HOM absorbers, and instrumentation wires. Unlike the concept for ILC and the design of XFEL, CW cryomodules typically have no thermal radiation shield at this temperature level since RF losses already provide 10s of Watts or more per cryomodule at the 2 K level.

The highest temperature level will be helium in the range 35–85 K or perhaps liquid nitrogen at around 80 K. This temperature level provides not only conductive thermal intercepts but also a thermal radiation shield. At many institutions, there will be no liquid nitrogen in the tunnel, implying the use of a gaseous helium circuit. However, often for test purposes in various test cryostats and facilities, as well as in some accelerators, the "70 K" thermal shield may be cooled with liquid nitrogen at approximately 80 K. This higher temperature will have some impact on thermal loads at the lower temperatures which should be assessed.

The thermal shield must be designed such that introduction of cold (process temperature) helium into the thermal shield piping when the thermal shield is warm, resulting in a very fast cool-down, does not damage the thermal shield or other parts of the cryomodule. (The issues are warping and associated forces, thermal stresses, etc.) Slots cut into the thermal shield perpendicular to the trace piping allow thermal contraction of the pipe ahead of thermal shield cooling. Thermal shield trace piping is arranged such that counterflow heat transfer does not inhibit cool-down of the thermal shield.

Heat loads for early TESLA-style cryomodules are summarized in Sect. 5.8, below. Special considerations for the high heat loads at 2 K with CW operation are described in Sect. 5.6.2 of this document.

Evacuated multilayer insulation (MLI) is used within the cryomodule on the thermal radiation shield, piping, and helium vessels. MLI on colder piping and vessels under the thermal radiation shield, while not very effective in terms of reducing overall heat load, greatly reduces boiloff rates from loss of vacuum incidents, in turn reducing emergency venting pipe and valve size requirements.

Each helium vessel includes an electric heater for 2 K flow and pressure control. In order to avoid cold feed-throughs from superfluid helium to insulating vacuum, these heaters are installed on the outsides of the helium vessels. The presence of a steady-state pressure drop results in a pressure change at the cryomodule with a change in flow rate (e.g. due to heat load change or liquid level control valve position change), even with constant cold compressor inlet pressure (perfect cryoplant pressure regulation). Heaters distributed within the cryomodules are required to compensate for heat load changes so as to control subsequent flow and pressure changes.

Cables are thermally intercepted at the 70 K level. Special attention is given to thermal intercepting of the piezo actuator wires and housing so as to assure piezo temperatures remain below 80 K, to improve lifetime and performance. Input coupler thermal intercepts are provided at 5 K and at 70 K.

Figure 5.12 illustrates the cryomodule cross-section and cryogenic circuit labels. Figure 5.13 shows the cryomodule flow scheme. The large-scale cryogenic system concept for TESLA is described in [11, 12].

Line A, the 2.2 K supply line, provides helium above the critical pressure (above 2.2 bar), so single-phase flow at all temperatures, to the liquid level control valves which maintain the saturated 2 K liquid level in the 2-phase pipe over the RF cavities.

Lines C and D are the low temperature helium thermal intercept supply and return lines. Lines E and F, similarly, are the higher temperature thermal intercept (and thermal shield) supply and return lines.

Line H, the warm-up/cool-down line, provides helium to capillary tubes which supply the bottom of each helium vessel, allowing parallel warm-up or cool-down of the string of helium vessels.

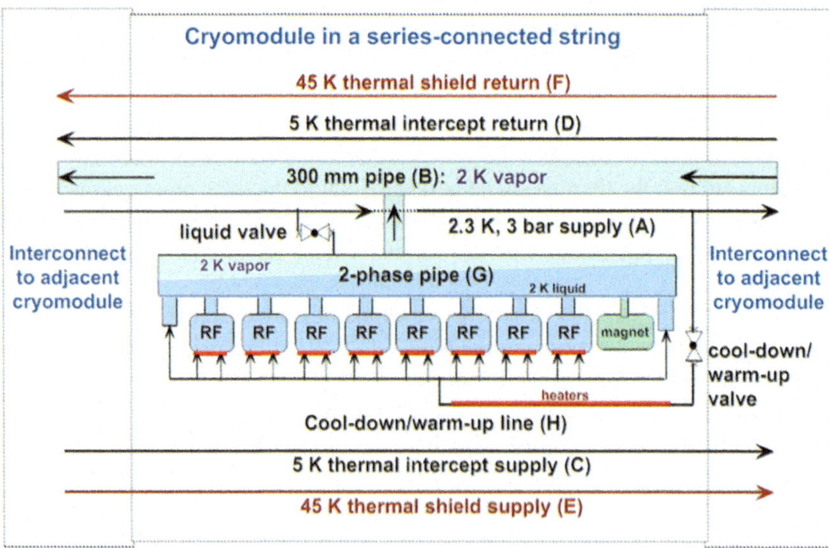


**Fig. 5.13** LCLS-II 1.3 GHz cryomodule schematic. For ILC, the 2-phase pipes and cool-down pipes are connected in series through a string of cryomodules

Lines passing through cryomodule interconnects are all welded, the only mechanical seals being the vacuum O-rings and all-metal beam tube vacuum seals.

### 5.6.2 *Design for Large 2 K Heat Transport and Helium Flow*

Heat from the outside surface of the niobium RF cavity, and heat entering via conduction from the beam pipe at the RF cavity ends, is carried through stagnant saturated Helium II to the liquid helium surface in the 2-phase pipe via superfluid heat transport. For heat transport through saturated superfluid helium around 2.0 K, 1 W/cm$^2$. is a conservative rule for a vertical pipe [13, 14].

The critical heat flux for a non-vertical pipe connection from the helium vessel to the 2-phase pipe may be considerably less than 1 W/cm$^2$ (Fig. 5.14). Configurations other than vertical require analysis to verify that the anticipated heat flux is less than the critical heat flux. Also, temperatures above 2.0 K result in a lower critical heat flux due to reduced superfluid heat transport. For LCLS-II, these considerations have resulted in our increasing the inner diameter of the nozzle (or "chimney") from the helium vessel to the 2-phase pipe from 55 mm (the TESLA/XFEL size) to 95 mm.

End flange to cavity including support tabs must have sufficient helium cross-sectional area to carry heat from the cavity ends and beam tube into the bulk helium in the helium vessel (Fig. 5.15). Arrows indicate conduction through end of power coupler outer conductor and beam pipe followed by heat transport through a thin layer of liquid helium at the end of the first cavity cell.)

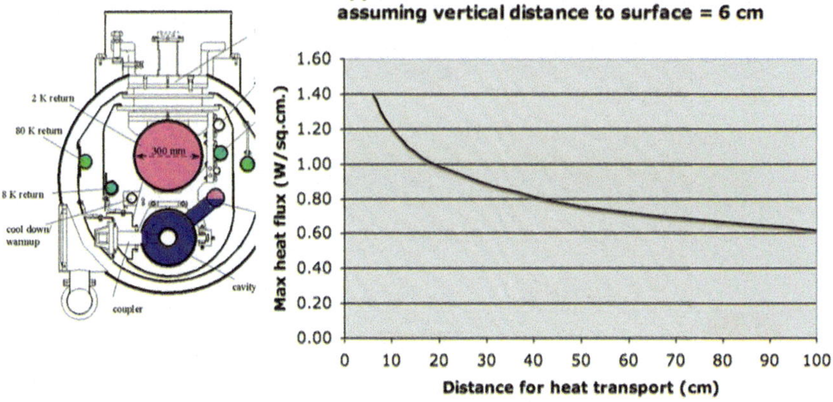


**Fig. 5.14** Maximum heat flux in saturated superfluid at 2.0 K

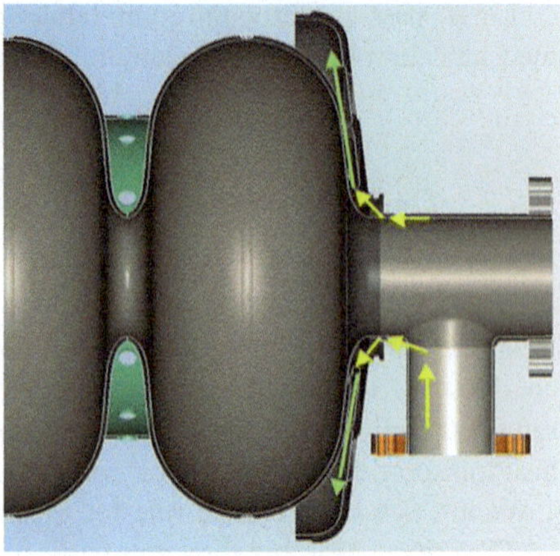

**Fig. 5.15** Heat transport path from input coupler and beam tube region to 2 K helium

Evaporation from the surface of the saturated helium liquid results in vapor flow within the 2-phase pipe over the liquid surface to an exit port connecting to the HGRP. Studies done at CEA Grenoble for CERN's LHC magnet cooling [15] provided the result that a 5 m/s vapor "speed limit" over liquid is a conservative "rule of thumb" not to entrain liquid droplets into the vapor. For LCLS-II, the closure of each 2-phase pipe in each cryomodule, limiting the 2-phase pipe vapor flow to that generated by one cryomodule, keeps the vapor flow rate quite low, less than 2 m/s. Nevertheless, the 0.5 % slope (Fig. 5.16) combined with the fact that the connecting nozzle from the helium vessel is 100 mm OD results in a 100 mm OD 2-phase pipe, larger than the 69 mm for XFEL. The result is that the pressure drop within the 2-phase pipe is negligibly small (Fig. 5.17).

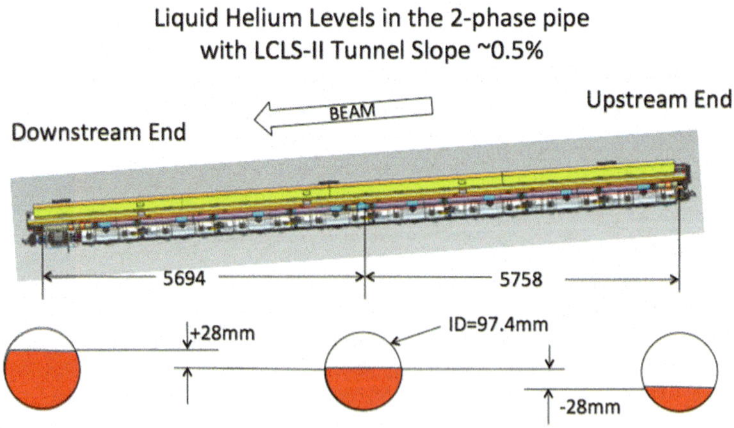


**Fig. 5.16** Liquid level illustration for single LCLS-II cryomodule (*credit* for figure—Fermilab)

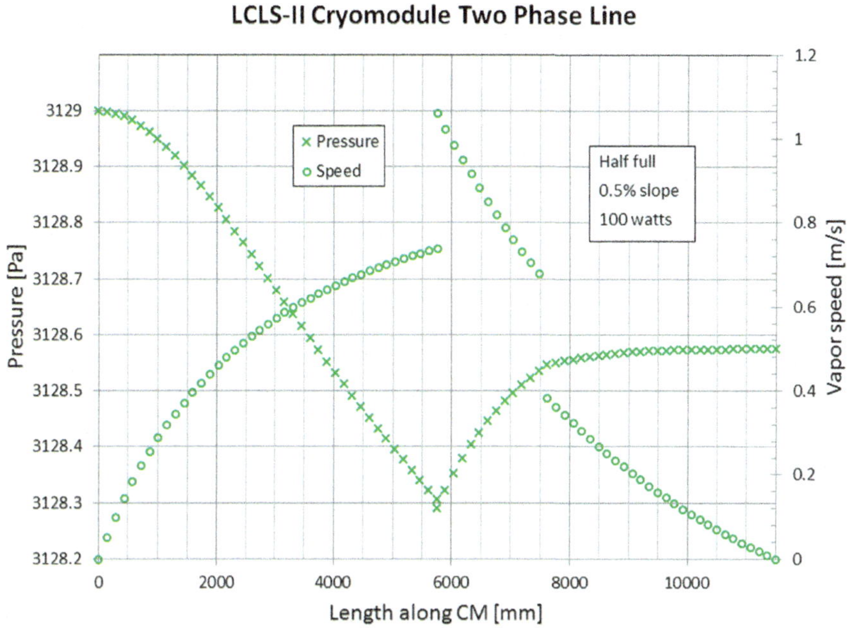


**Fig. 5.17** Pressure drop and vapor velocity in 2-phase pipe in a single cryomodule. The vapor exits from the 2-phase pipe at the 5700 mm central location. Liquid helium is added, with some vapor also, at the 7500 mm location. The asymmetry of vapor speed and pressure drop is due both to the off-center input of the 2-phase flow and a 0.5 % upward slope from *left to right* in the figure (*credit* for analysis and plot—Joshua Kaluzny, Fermilab, and Joel Fuerst, Argonne National Laboratory)

### *5.6.3 Pressure Drop Analyses*

Pressure drops within and through the cryomodule must be analyzed in combination with the helium distribution system. Pipes are sized for the worst case among steady-state, peak flow rates, upset, cool-down, warm-up, and venting and conditions. Pressure drops must be analyzed for each helium flow path to ensure that steady-state operation matches system design and that non-steady conditions (cool-down, emergency venting, warm-up) are properly handled. Input variables for these analyses include line size, allowable temperature rise, allowable pressure drop, and heat load. (Temperature rise and heat load determine mass flow for supercritical pressure helium or helium gas in the thermal intercept pipes and 2 K supply pipe.)

For the HGRP, pressure drop is particularly important and limited, since this pressure drop determines the helium temperature at the helium vessel, hence RF cavity temperature. We design for a pressure drop <10 % of total pressure in normal operation.

Maximum allowable pressure for emergency venting, combined with distances to relief devices may also influence line sizes [16]. Helium piping and vessels vent

**Table 5.5** Cryomodule pipe size comparison

| Pipe function | Pipe name (Fig. 5.12) | TTF inner diameter (mm) | XFEL inner diameter (mm) | ILC inner diameter (mm) | LCLS-II inner diameter (mm) |
|---|---|---|---|---|---|
| 2.2 K subcooled supply | A | 45.2 | 45.2 | 60.2 | 54.8 |
| Gas helium return header, structural support | B | 300 | 300 | 300 | 300 |
| 5 K shield and intercept supply | C | 54 | 54 | 56.1 | 54.8 |
| 5 K shield and intercept return | D | 50 | 65 | 69.9 | 50.8 |
| High temperature shield and intercept supply | E | 54 | 65 | 72.0 | 54.8 |
| High temperature shield and intercept return | F | 50 | 65 | 79.4 | 52.5 |
| 2-phase pipe | G | 72.1 | >72.1 | 69.0 | 97.4 |
| Helium vessel to 2-phase pipe nozzle ("chimney") | | 54.9 | 54.9 | 54.9 | 95 |
| Warm-up/cool-down line | H | | | 38.9 | 38.9 |

into the adjacent cryomodules and out to the distribution system, allowing placement of all process relief valves in the distribution system.

Loss of vacuum venting: pressure in the helium vessel of the dressed cavity less than the cold maximum allowable working pressure (MAWP) of the helium vessel and dressed cavity. Venting path includes nozzle from helium vessel, 2-phase pipe, may include gas return pipe, and also includes any external vent lines. Worst-case heat flux to liquid helium temperature metal surfaces with loss of vacuum to air is assumed to be 4.0 W/cm$^2$. Worst-case heat flux to liquid helium temperature surfaces covered by at least 5 layers of multilayer insulation (MLI) is assumed to be 0.6 W/cm$^2$.

Finally, we match cryomodule and cryogenic distribution system design to the cryogenic plant in terms of providing flow rates, temperatures, and pressures consistent with cryogenic plant requirements (Table 5.5).

### 5.6.4 *Typical TESLA-Style Cryomodule Maximum Allowable Working Pressures*

See Table 5.6.

**Table 5.6** Maximum allowable working pressures for typical TESLA-style cryomodules

| Region | Warm MAWP (bar) | Cold MAWP (bar) |
|---|---|---|
| 2 K, low pressure space | 2.05 | 4.1 |
| 2 K, positive pressure piping (separated by valves from low P space) | 20.0 | 20.0 |
| 5 K piping | 20.0 | 20.0 |
| 45 K piping | 20.0 | 20.0 |
| Insulating vacuum space | 1 atm external with full vacuum inside 0.5 positive differential internal | |
| Cavity vacuum | 2.05 bar external with full vacuum inside 0.5 positive differential internal | 4.1 bar external with full vacuum inside 0.5 positive differential internal |
| Beam pipe vacuum outside of cavities | 1 atm external with full vacuum inside 0.5 positive differential internal | 1 atm external with full vacuum inside 0.5 positive differential internal |

### 5.6.5 *Instrumentation*

The cryomodule must be instrumented with liquid level probe (or probes) for the 2-phase helium II system when level control is incorporated into each cryomodule, thermometry for cool-down and monitoring of critical input coupler, HOM coupler, and current lead temperatures, and other instrumentation. Table 5.7 lists possible instrumentation for a CW electron linac cryomodule.

### 5.6.6 *Cryomodule Test Requirements*

At least some fraction of cryomodules are typically tested prior to installation for the following:

- Leak and pressure tests for quality assurance and pressure code compliance.
- Temperature profiles
- Approximate heat loads
- RF cavity performance
- Tuner performance
- Instrumentation

### 5.6.7 *Pressure Stability at the 2 K Level*

It is possible to generate pressure pulses within a cryomodule, for example via heat input from the warm end of a closed pipe. Hence, one should try to avoid

**Table 5.7** Example of cryomodule instrumentation for a CW electron linac cryomodule (LCLS-II). Production cryomodules may have a reduced set of necessary instrumentation

| Sensor description | Quantity and location |
|---|---|
| Piezo actuators (Fast Tuners) | Typically 2 or 4 per tuner |
| Stepper motors | One stepper motor per tuner |
| Embedded platinum temperature sensors on the stepper motors | One for each stepper motor |
| Cavity field probes—transmitted power | 1 field probe per cavity |
| HOM field probes—transmitted power | 1 on each HOM × 2 HOMs per cavity |
| Coupler Electron Pick-ups | 2 Pick-ups per cavity (1 is mounted on the cavity side of the coupler and 1 is mounted on the wave guide side of the coupler). |
| Helium vessel cernox RTDs | 1 RTD at each cavity |
| Main coupler platinum RTDs | 2 RTDs at each cavity (1 at 12 o'clock & 1 at 10 o'clock). |
| Beam position monitor (BPM) | 1 BPM per cryomodule—4 pick-ups |
| Coupler tuner motors | 1 stepper motor, 2 limit switches, and 1 potentiometers for each main coupler (if remote coupler tuning is incorporated) |
| Cool down temperature sensors | 2 at the bottom of the 5 K shield and 3 inside the 300 mm pipe at each end |
| Cool down temperature sensors | 2 RTDs—mounted on the bottom of the 70 K |
| HOM coupler RTDs | 2 RTDs on each HOM coupler |
| BPM Pick-Up Temperature Platinum RTDs | 2 platinum RTDs—1 primary and 1 redundant |
| Quadrupole and corrector coil voltage taps for coil quench protection | 6 VT's (3 Pairs)—2 VT's for the quad, 2 Vt's for each of 2 correctors |
| Quadrupole and corrector coil cold mass temperature sensors | 1 RTD for each coil package × 4 coil packages |
| Quadrupole and corrector coil wire heater | 1 wire heater—1 heater for the entire coil package |
| Quadrupole and corrector coil voltage taps for power lead quench or current protection | 2 voltage taps/current lead × 2 leads/coil |
| Quadrupole and corrector coil lead RTDs | 2 at 4 K region, 2 on 4 K intercept, 1 on 4 k-70 K region, and 1 at 70 K intercept |
| Helium vessel heaters | 1 electric heater per helium vessel |
| Magnetic sensors | Optional, typically for R&D or prototype |
| Beam loss monitor | Depends on linac requirements |
| Liquid level sensors | If cryomodules have individual liquid levels |

"dead-headed" lines which can warm up, for example, the line terminating at the cool-down valve after it is closed. To avoid a warm valve providing such a warm termination on the closed pipe, one solution is to locate the valve lower than the supply pipe such that cold helium sits on the valve.

The relatively large vapor volume contained in the HGRP combined with the use of helium II result in an inherently stable pressure, since pressure fluctuations are not only damped by the large vapor volume, but helium II heat transport supports liquid-vapor equilibrium at saturation temperature. An advantage of helium II, aside from the tremendous heat transport capability, is this inherent pressure stability, which in turn reduces the problem of helium pressure fluctuations affecting cavity resonant frequency.

## 5.7 Cryomodule Helium Inventory

Table 5.8 lists cryomodule helium inventory for the various lines and volumes. Note that most of the helium mass is contained in the helium vessels, the liquid around the RF cavities. Depending on how liquid level is managed, there may also be additional liquid helium in the 2-phase pipe. The large volume of the 300 mm pipe, filled with low-pressure helium vapor, provides a nice buffer to damp pressure fluctuations and to receive helium gas with sudden boiling due to some off-normal condition.

## 5.8 Early Results from the TESLA Cryomodules

The ILC cryomodule design evolved from the cryomodules designed for the TESLA (TeV Superconducting Linear Accelerator) project in the 1990s. [1] Early experimental results from these cryomodules led to design improvements seen in the ILC cryomodules and also illustrate several important features of cryostat design.

Table 5.9 shows the predicted and measured static heat leak for first two TESLA cryomodules built and tested [17].

Notice that the measured heat leak of cryomodule #1 significantly exceeds the predicted static at all three temperature levels with the 2 K heat load being more than twice the predicted value. The explanation of a significant part of this discrepancy can be seen in Fig. 5.18. Look first at the upper part of the figure. This shows the measured temperatures at various points on both the 70 and 5 K thermal shields and the support posts. Notice on the right side of the figure that sensors on the innermost (5 K) thermal shield indicate temperature of 13–17 K rather 6–7 K as shown on the rest of the shield. This result is indicative of an increased heat load to that shield in that location. Cryomodule #1 was instrumented with a stretched wire measurement system that allowed real time monitoring of the position of the superconducting cavities so that the support system could be verified to meet the alignment requirements (see below). This measurement system included large numbers of instrumentation wire connecting the 2 K system to room temperature. In cryomodule #1, there was inadequate heat sinking of the wires at the 70 K level and thus much more heat than predicted was deposited at the 5 K shield level and at the 2 K level. There was also found to be issues with the proper treatment of the MLI penetrations (see Chap. 1) at the power coupler ports; which also added to the static heat leak.

**Table 5.8** ILC cryomodule line sizes, helium conditions, and helium inventory

| Location | State | Temp (K) | Pressure (bar) | Pipe ID (mm) | Density (kg/m$^3$) | Volume (L) | Mass (kg) | Equiv liquid (L) |
|---|---|---|---|---|---|---|---|---|
| One helium vessel | Liquid | 2.00 | 0.031 | | 145.700 | 23.0 | 3.35 | 26.8 |
| 9 helium vessels | Liquid | 2.00 | 0.031 | | 145.700 | 207.0 | 30.16 | 241.3 |
| 2-phase pipe | Vapor | 2.00 | 0.031 | 69.0 | 0.830 | 47.3 | 0.04 | 0.3 |
| 2 K supply pipe | Subcooled | 2.40 | 1.200 | 60.2 | 147.800 | 36.0 | 5.32 | 42.6 |
| 300 mm pipe | Vapor | 2.00 | 0.031 | 300.0 | 0.830 | 893.9 | 0.74 | 5.9 |
| 5 K supply | Supercritical | 5.00 | 5.000 | 56.1 | 129.000 | 31.3 | 4.03 | 32.3 |
| 8 K return | Supercritical | 8.00 | 4.000 | 69.9 | 30.760 | 48.5 | 1.49 | 11.9 |
| 40 K supply | Gas | 40.00 | 16.000 | 72.0 | 18.460 | 51.5 | 0.95 | 7.6 |
| 80 K return | Gas | 80.00 | 14.000 | 79.4 | 8.220 | 62.6 | 0.51 | 4.1 |
| Warmup line | Vapor | 2.00 | 0.03 | 38.9 | 0.830 | 15.0 | 0.01 | 0.1 |
| Totals | | | | | | | | |
| One module | | | | | | 1393 | 43 | 346 |
| Short string | 9 modules | | | | | 12,538 | 389 | 3115 |
| Avg cryo unit | 21 short string (system on one cryogenic plant) | | | | | 263,288 | 8177 | 65,414 |

**Table 5.9** Predicted and measured static heat leak for TESLA cryomodules #1 and #2 [17]

| Temperature level (K) | Predicted heat leak (W) | Measured heat leak (W) cryomodule #1 (alone) | Measured heat leak (W) cryomodule #1 (with #2) | Measured heat leak (W) cryomodule #2 |
|---|---|---|---|---|
| 70 | 76.8 | 90 | 81.5 | 77.9 |
| 4.5 | 13.9 | 23 | 15.9 | 13 |
| 2 | 2.8 | 6 | 5 | 4 |

Prior to the later test of cryomodule #1 with cryomodule #2, some of the issues above were addressed and as a result the heat leak performance of Cryomodule #1 improved somewhat. Cryomodule #2 was constructed with far fewer instrumentation wires associated with the stretched wire system and those that remained were better heat sunk. Examining the lower part of Fig. 5.18, which shows the shield temperature distribution of cryomodule #2 shows the impact of these changes. The temperatures at the right side of the 5 K thermal shield have been reduced and are much closer to the nominal 5 K level. This result is also reflected in Table 5.9. The measured heat leak for cryomodule #2 is within predictions for the 4.5 K level and much closer to predictions for the 70 K and 2 K levels. Based on these experiences, later ILC cryomodules have been built that meet the predicted static heat loads [18]

The story above illustrates the importance, as described in Chap. 1, of careful design and heat sinking of instrumentation wires and of the need for proper installation of MLI in penetrations.

The stretched wire system [19] that caused early heat leak problems did however, provide very useful data on the performance of the support system. Figure 21 shows the results of the displacement of the wire position monitors (WPMs) of the stretched wire systems between room temperature and cryogenic temperatures for cryomodule #2.

The various WPM positions are correlated to various cavity and magnet locations wthin the cryomodule. WPM positions 1–10 represent SRF cavity positions while positions 11 and 12 represent the superconducting magnet assembly in the cryomodule. Notice from Fig. 5.19 that the maximum displacement occurs at the end of the cryomodule. This is a result of unbalanced forces on and resultant bending of the 300 mm gas return pipe which acts as the structural backbone of the cryomodule at to which the cavities and magnet assembly are fixed.

The allowed displacement for the cavities from their ideal location when cold is ±0.5 mm. Note that the measurements show that all the cavities, even the first one are within this specification. However, for the superconducting magnet assembly, the result is different. The magnet package is required to be within ±0.25 mm of its ideal location when cold. The data in Fig. 5.19 shows that magnet assembly located at the end of the cryomodule does not meet this requirement. As a result of these measurements, changes have been made in the later XFEL and ILC cryomodules. The support post at the magnet end of the cryomodule has been moved closer to the magnet location. This has the effect of stiffening the 300 mm pipe near the magnet

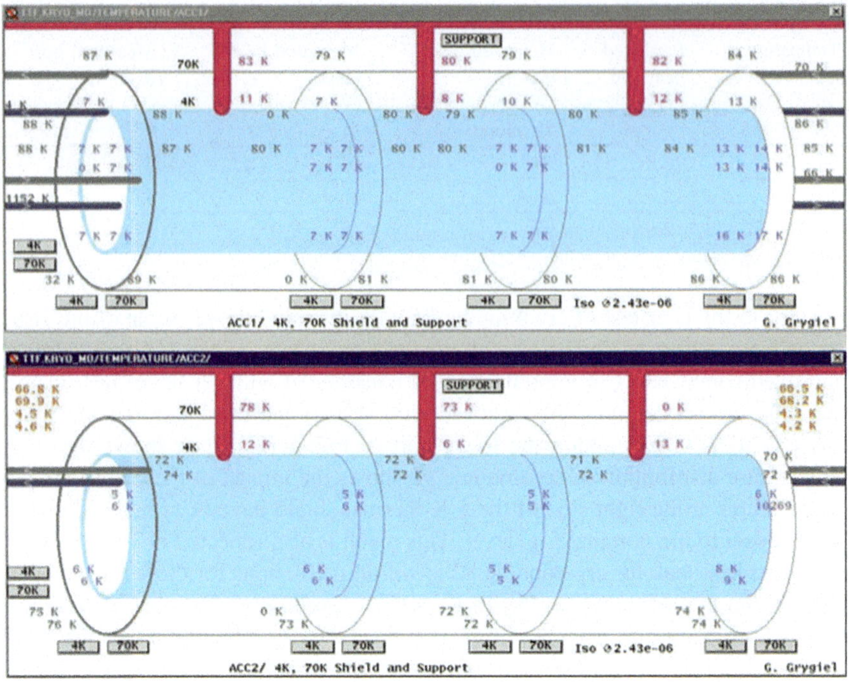


**Fig. 5.18** Temperature measurements of the thermal shields and support posts for TESLA cryomodule #1 & cryomodule 2 [17]

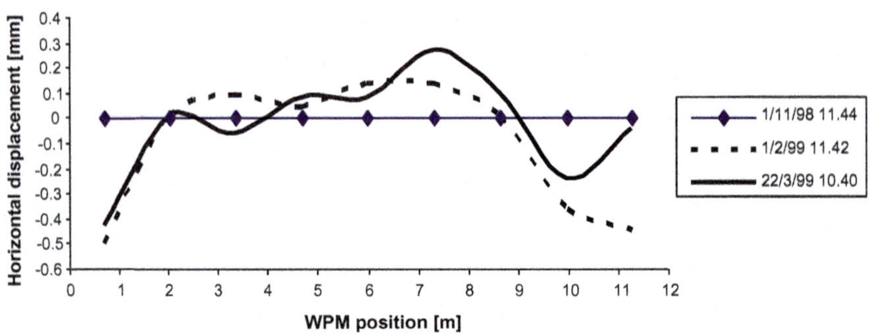


**Fig. 5.19** WPM displacement between room temperature (1/11/98) and 2 K for two separate cool downs (1/2/99 and 22/3/99) [17]

assembly and reduces the displacement of the magnet assembly to a limit suitable for use in XFEL and LCLS II. In the case of the much larger ILC machine, which has tighter alignment requirements, the final design change will likely be to move the magnet assembly to the middle of the cryomodule where the displacement is minimized.

In addition to illustrating important aspects of cryostat design, the early experiences with the TESLA cryomodules also shows the importance of prototype tests in cryostat design. The construction and testing of a prototype cryostat and the subsequent altering of the design based on the prototype results, is an important part of a successful cryostat project. Large systems such as the SSC magnets (Chap. 2), LHC magnets (Chap. 3), SHOOT (Chap. 4), ILC (Chap. 5) and the 12 GeV Upgrade (Chap. 6) all take this approach. It is important to allow sufficient time in any project schedule for the prototyping process.

## 5.9 Modifications for CW Operation in the LCLS-II Linac

Modifications of a TESLA-style cryomodule for LCLS-II illustrate design features for CW operation and design for relatively larger heat loads than a pulsed linac would typically have. Similar considerations for high heat loads are described for CW operation in Energy Recovery Linacs (ERL) by Kugeler et al. [14] LCLS-II adopted the most recent 8-cavity TESLA-style as developed for XFEL, with magnet at the end (unlike the ILC concept, which moved the magnet to the center), but modified for CW operation and for the SLAC tunnel 0.5 % slope. Some major differences from the TESLA-style design include:

- Modified power input coupler for CW power,
- Larger diameter 2-phase 2 K helium piping for increased heat and vapor transport and for liquid helium management with the tunnel slope,
- Closure of each end of the 2-phase pipe with the addition of a liquid helium supply valve in the cryomodule,
- Closure of each end of the cool-down/warm-up pipe manifold for individual cryomodule cooling to facilitate Q0 preservation, with the addition of a cool-down valve in each cryomodule,
- New end lever tuner design for increased stiffness and fine control,
- The addition of tuner access ports to facilitate inspections or small repairs of tuner components,
- Removal of the 5 K thermal shield due to its marginal benefit relative to dynamic heat loads and the advantages of simplification of design,
- And enhanced magnetic shielding for Q0 preservation.

## 5.10 Summary

The TESLA collaboration developed a unique variant of SRF cryomodule designs, the chief feature being use of the large, low pressure helium vapor return pipe as the structural support backbone of the cryomodule. Additional innovative features include all cryogenic piping within the cryomodule (no parallel external cryogenic transfer line), long strings of RF cavities within a single cryomodule, and cryomodules connected in series. Several projects, including FLASH and XFEL at DESY, LCLS-II at SLAC, and the ILC technical design have adopted this general design concept.

Advantages include saving space by eliminating the external transfer line, relatively tight packing of RF cavities along the beamline due to fewer warm-cold transitions, and potentially lower costs. However, a primary disadvantage is the relative lack of independence for warm-up, replacement, and cool-down of individual cryomodules.

**Acknowledgments** We thank the scientists, engineers, and designers in the TESLA, ILC, and LCLS-II collaborations, and in the TESLA Technology Collaboration for the use of figures from 3-D cryomodule design models and other information about TESLA-style cryomodules.

## References

1. TESLA, Technical design report, part II: the accelerator, published by DESY, Notkestrasse 85, 22607 Hamburg, Mar 2001
2. International linear collider reference design report—ILC global design effort and world wide study, in *Volume 3: Accelerator*, eds. by N. Phinney, N. Toge, N. Walker (2007)
3. International linear collider technical design report—ILC global design effort and world wide study, in *Part II: The ILC Baseline Reference*, eds. by J. Carwardine, N. Phinney, N. Toge, P. Burrows (2012)
4. TESLA Test Facility Linac Design Report, Editor: Don Edwards, March 1995
5. The European X-ray free-electron laser—technical design report. DESY XFEL project group, European XFEL project team, Deutsches Elektronen-Synchrotron, Member of the Helmholtz Association, Notkestrasse 85, 22607 Hamburg, Germany, July 2007. http://www.xfel.net
6. LCLS-II Final Design Report (LCLSII-1.1-DR-0251-R0) November 22, 2015
7. T.H. Nicol, R.C. Niemann, J.D. Gonczy, SSC magnet cryostat suspension system design, in *Advances in Cryogenic Engineering*, vol. 33 (Plenum Press, New York, 19880, pp. 227–234
8. T.H. Nicol, R.C. Niemann, J.D. Gonczy, A suspension system for superconducting super collider magnets, in *Proceedings of the Eleventh International Cryogenic Engineering Conference* (Butterworth & Co., Surrey, UK, 1986), pp. 533–538
9. T.H. Nicol, TESLA test cell cryostat support post thermal and structural analysis, Fermilab TM-1794, Aug 1992
10. A. Ballarino, Conduction-cooled, 60 a resistive current leads for LHC dipole correctors, LHC project report 691, European laboratory for particle physics. CERN CH—1211 Geneva 23 Switzerland, 5 Mar 2004
11. G. Horlitz, T. Peterson, D. Trines, The TESLA 500 cryogenic system layout, in *Advances in Cryogenic Engineering 41A*, ed. by P. Kittel, et al. (Plenum, New York, 1996), pp. 911–920
12. S. Wolff, H. Lierl,B. Petersen,J. Weisend II, The cryogenic system of TESLA 500—An Update, in *Proceedings of the International Cryogenic Engineering Conference ICEC 17* (Bournemouth, U.K., 1998), pp. 879–882
13. T. Peterson, Fermilab, Notes about the limits of heat transport from a TESLA helium vessel with a nearly closed saturated bath of helium II", TESLA report #94-18 (June 1994)
14. O. Kugeler, A. Neumann, W. Anders, J. Knobloch, Adapting TESLA technology for future CW light sources using HoBiCaT. Rev. Sci. Instrum. **81**, 074701 (2010) (Helmholtz-Zentrum-Berlin (HZB), Berlin, Germany, p. 12489)

15. B. Rousset, A. Gauthier, L. Grimaud, R. van Weelderen, Latest developments on He II co-current two-phase flow studies, in *Advances in Cryogenic Engineering, Cryogenic Engineering Conference*, vol. 43B , (1997), pp. 1441–1448
16. Petersen, B., Wolf, S., Numerical simulations of possible fault conditions in the cryogenic operation of the TTF/FEL—and Tesla linear accelerator, in *Proceedings of the 18th International Cryogenic Engineering Conference* (*ICEC18*), Mumbai, India (2000)
17. C. Pagani et al., Construction, commissioning and cryogenic performances of the first TESLA test facility (TTF) cryomodule, Adv. Cryo. Eng. **43A** (1998)
18. N. Ohuchi et al., Thermal performance of the S1-Global cryomodule for the ILC, in *Proceedings of IPAC2011* (2011)
19. D. Giove, A wire position monitor (WPM) system to control the cold mass movements inside the TTF cryomodule, in *Proceedings of the 1997 Particle Accelerator Conference*, vol. 3 (1997)

# Chapter 6
# Segmented SRF Cryomodules

**E. Daly, Thomas H. Nicol and J. Preble**

**Abstract** Linear accelerators based on Superconducting Radio Frequency (SRF) cavities have become increasingly important and wide spread. This chapter reviews the design and operation of four different SRF segmented cryomodules. These cryomodules were used in projects from the 1980s to the 2010s and as such this chapter provides a good overview of the development of this type of cryostat. Topics described include: requirements, cooling of SRF cavities, tuners, power couplers, structural and thermal insulation systems, magnetic shields, vacuum systems and instrumentation. Performance results from both prototypes and series operation are presented.

## 6.1 Introduction

An important application of cryogenics in particle accelerators is the cooling of Superconducting Radiofrequency (SRF) cavities. These cavities provide acceleration to the charged particle beam by storing radiofrequency (RF) energy in a resonant structure (or cavity). The system is designed such that the electrical field produced in the SRF cavity accelerates the charged particles as they pass through the cavity. The cavities are made from pure niobium operated in its superconducting state to minimize the wall losses of the RF energy. Thus, more of the input RF energy goes into accelerating the beam itself.

E. Daly (✉) · J. Preble
Thomas Jefferson National Laboratory, 12000 Jefferson Avenue,
Newport News, VA 23606, USA
e-mail: edaly@jlab.org

J. Preble
e-mail: preble@jlab.org

T.H. Nicol
Fermi National Accelerator Laboratory, P.O. Box 500, Batavia
IL 60510, USA
e-mail: tnicol@fnal.gov



J.G. Weisend II (ed.), *Cryostat Design*, International Cryogenics Monograph Series,
DOI 10.1007/978-3-319-31150-0_6

Since this application involves RF energy and is thus an alternating current application of superconductivity, there are always going to be wall losses and heating within the superconducting niobium. These losses are a function of a number of parameters including the RF frequency and the temperature. The optimal temperature for most SRF applications is between 1.8 and 2.1 K.

The SRF cavities are built into cryostats known as cryomodules. These are described below and in Chap. 5. In addition to the cavities themselves, SRF cryomodules contain other components. Power couplers (or couplers) bring the RF energy from the room temperature power supplies to the SRF cavities. Tuners adjust the resonant frequency of the cavities by physically changing the cavities shape. Higher order mode (HOM) couplers remove unwanted frequencies generated by the beam moving through the SRF cavity. A complementary approach is to absorb the HOM frequencies in a dissipative material to prevent its propagation. Such materials are known as HOM absorbers or HOM loads. In addition, the cryomodules typically contain magnetic shielding to protect the cavities from stray magnetic fields (including the earth's) which can degrade cavity performance.

Requirements for cryomodules include: low heat leak, low vibration, tight alignment tolerances on the cavities and a high vacuum level in the beam tube vacuum for optimum cavity performance.

There are two ways that cryogenic and vacuum systems are distributed to individual modules in superconducting magnets or cavity strings. The first, referred to as fine segmentation or segmented, refers to systems in which the insulating vacuum and the cryogenic circuits are confined to an individual cryomodule or cryostat with the only connection between modules being the beam tube. This approach has a number of advantages; it may allow for individual warm up and cool down of cryomodules, the separate cryostat isolation vacuums prevents a failure in one isolation vacuum space from propagating to other cryostats and the space in between the segmented cryomodules may be filled with magnets, beam instrumentation and vacuum systems that are more easily operated at 300 K. Fine segmentation is the most common approach used in SRF cryomodules and is seen in the examples given in this chapter as well as in the European Spallation Source (ESS) [1] and Facility for Rare Isotope Beams (FRIB) [2] projects.

There are disadvantages to fine segmentation. The biggest is that the many warm to cold transitions add additional heat leak and are quite expensive. Thus, for systems involving large numbers of cryostats or cryomodules, the second approach sometimes called coarse segmentation or continuous is used. Course segmentation refers to systems in which the cryogenic circuits and insulating vacuum inside individual cryostats are more or less continuous for long lengths, at least over the length of several cryomodules. Accelerator magnet systems such as the SSC (Chap. 2) and the LHC (Chap. 3) are configured this way as are the SRF cryomodules envisioned for the ILC, LCLS II and the XFEL at DESY (Chap. 5).

Generally speaking, systems involving less than 100 cryostats or cryomodules use the segmented approach while those involving more than 100 components use the continuous approach.

This chapter describes the design and experience of four different segmented cryomodules. The cryomodules are described in rough chronological order of their design and were designed between the 1980s and the 2010s. Since these cryomodules have many similar requirements, their design solutions are similar. However, an evolution of design can be also seen over the course of the development of these cryomodules. Due to the historical nature of these descriptions, the units used as those in place at the time of the original designs. Thus, a mixture of SI and English units will be seen.

## 6.2 C20 Cryomodule Design for CEBAF

### 6.2.1 Introduction

The Continuous Electron Beam Accelerator Facility (CEBAF) was built as a 4 GeV, 200 μA, electron accelerating facility [3]. The C20 cryomodule is the original design for the CEBAF accelerator and was developed in the late 1980's. CEBAF was the world's largest installation of Superconducting Radio Frequency (SRF) accelerating cavities. CEBAF had 338 identical SRF cavities housed in 43 cryostats. The cavities were assembled in pairs which became the natural design segment for the cryomodules. Four of these pairs were housed in identical cryostats that would be assembled together along with end cans to form a single cryomodule (see Fig. 6.1). The exception to this was the first cryostat, a quarter cryomodule, which included a single cavity pair cryostat with end caps.

### 6.2.2 Modularity and Segmentation

The CEBAF accelerator (see Fig. 6.2) includes two antiparallel linacs, an injector, and three experimental halls. Each linac contains 25 repeating zones. The first 20 of these zones contain one C20 cryomodule each. A zone is 10 m long and contains a cryomodule and a warm beamline girder which are 8.35 and 1.65 m long respectively. Each cryomodule has two helium supplies and returns, a primary circuit operating at 2 K and a shield circuit operating at 50 K, eight SRF cavities with 5 kW radio frequency (RF) power feeds, and various electrical connections for RF

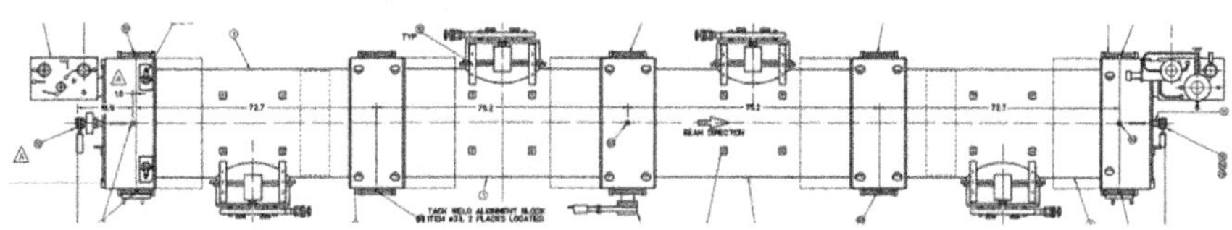

**Fig. 6.1** C20 CEBAF cryomodule

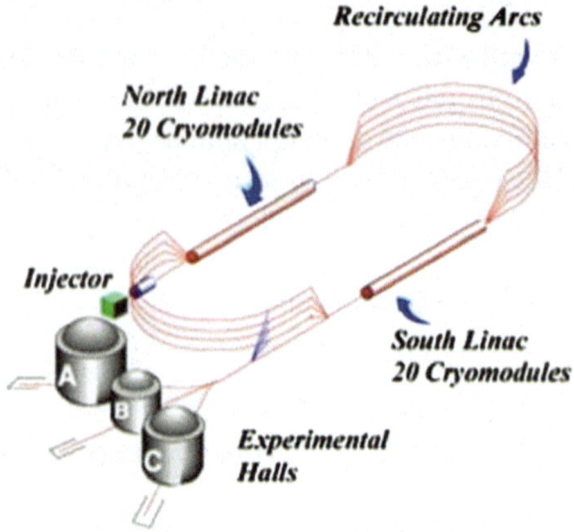


**Fig. 6.2** CEBAF accelerator

signals, instrumentation, and motor controls for the helium supply valve and cavity tuner stepper motors. A cryomodule can be removed and replaced with another cryomodule or drift tube in about a week minimizing accelerator operational down time. This has been valuable over the life of the machine as eleven cryomodules have been removed and refurbished over the first 23 years of operation.

## *6.2.3 Requirements*

The cryomodule requirements were detailed in a design handbook that covered the entire CEBAF accelerator ensuring appropriate system integration. The top level cryomodule requirements include an operating voltage of 20 MV, capable of accelerating 200 μA of beam recirculated five times, a nominal operating heat load of 68 and 140 W for the primary and shield circuits respectively. The eight SRF cavities contained in a cryomodule are required to be aligned relative to the nominal beam trajectory with a root-mean-square angular position of 2 mR and centered on the beam axis to ±0.02 in. The detailed requirements contained in the design handbook are the extension of these high level requirements to individual systems or components. Table 6.1 shows a summary of these requirements.

## *6.2.4 Design Description and Choices*

The original CEBAF cryomodule was the first large scale installation of SRF technology in the US. The adoption of this new technology was a risk and many design choices were made to allow for testing and verification of the cavity

**Table 6.1** Summary of CEBAF C20 cryomodule requirements from CEBAF design handbook (1991 revision)

| Item | Value | Units |
|---|---|---|
| Number of cryomodules | 42.25 | |
| Number of such sections used in injector | 2.25 | |
| Number of such sections used in linac | 40 | |
| Length of section, including magnetic elements | 9.60 | m |
| Number of cryounits per cryomodule | 4 | |
| Number of cavities per cryomodule | 8 | |
| *Cavities* | | |
| Type | Superconducting | |
| Duty cycle | CW | |
| Operating frequency | 1497, ±0.00002 | MHz |
| Accelerating gradient, active length | ≥5 | MV/m |
| Intrinsic limitation of accelerating gradient | ≥20 | MV/m |
| Power coupled into beam, per cavity | 2.5 | kW |
| Total linac-associated power consumption | ≤10 | MW |
| HOM impedances, for most important modes | 500, 170,000 | |
| Regenerative beam breakup threshold, per beamlet | >200 | Actual μA |
| Regenerative beam breakup threshold, per beamlet | >2000 | Computed μA |
| Current transport capability | 1000 | μA |
| Number of cells per cavity | 5 | |
| Operating mode | π | |
| Cell shape | Elliptical | |
| Fundamental power coupler, waveguide | Beamline | |
| HOM coupler | Beamline | |
| Output waveguide cutoff frequency | 1900 | MHz |
| Number of output waveguides | 2 | |
| Angle between waveguides | 90 | |
| Elastic pressure sensitivity | <60 | Hz/torr |
| Microphonic tolerance, cryomodule surface | 10 | microns/sec |
| *Cryostat (cryounit)* | | |
| Thermal intercept shield | 1 | |
| Magnetic shield | 2 | |
| Liquid Level probe | 1 | |
| Superinsulation blankets | 2 | |
| Support rods | 10 | |
| Heater (100 W in liquid bath) | 1 | |
| Cryogenic thermometers (minimum) | 2 | |
| *Bridge piece set* | | |
| Bridging ring, insulating vacuum | 1 | |
| Thermal shield closure with braided hose | 1 | |
| Superinsulation blankets | 2 | |
| Magnetic shield | 1 | |
| Beampipe with pump out | 1 | |
| Helium pumping line, with bellows | 1 | |

performance during assembly and to minimize potential for compromising the demonstrated performance by subsequent processes. Two SRF cavities were assembled into a hermetically sealed Cavity Pair (CP), see Fig. 6.3, which undergoes performance testing in a Vertical Test Area (VTA) [4] and are never vented after testing to avoid contamination.

The cavities have frequency tuners installed and this assembly is mounted in a 316L stainless steel helium vessel. The helium vessel is assembled into a cryostat that includes cold and warm layers of magnetic shielding, an intermediate temperature thermal shield, multilayer insulation and a vacuum vessel. Together this assembly is a Cryounit (CU) see Fig. 6.4. Four of these CUs are assembled with a supply and return cryogenic end can to make a Cryomodule (CM). The cryomodule is the smallest assembly that is transported to and installed in the accelerator.

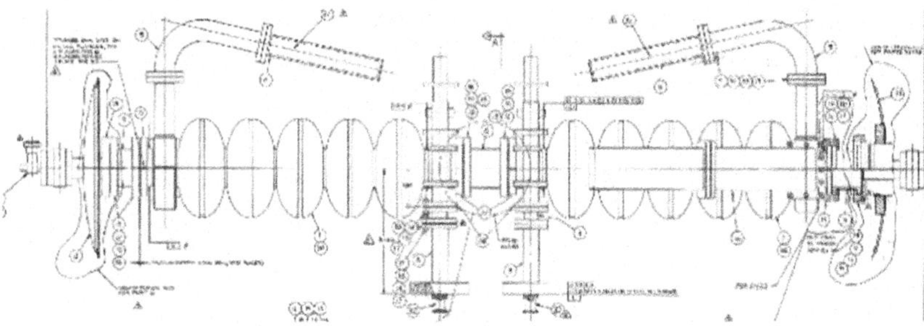

**Fig. 6.3** C20 cavity pair

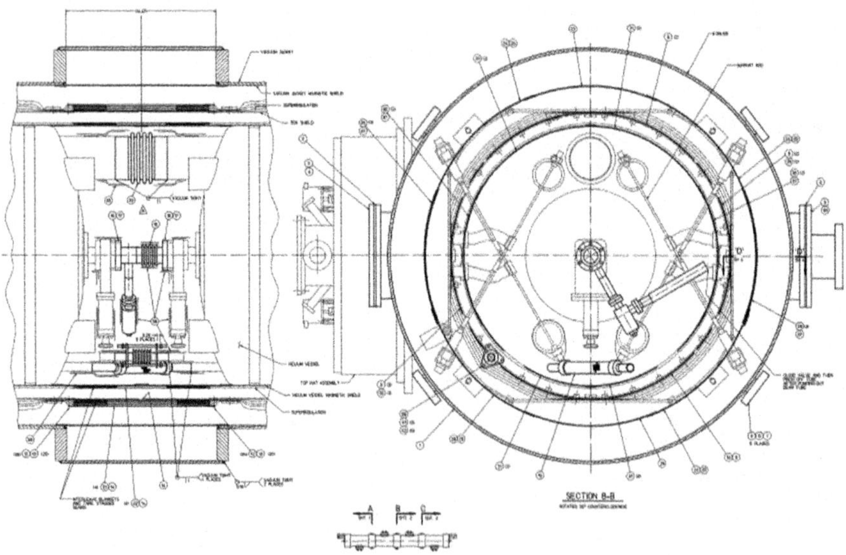


**Fig. 6.4** Bridging area between two cryounits

### 6.2.5 Cryogenic System Interfaces

Helium is supplied from a large central helium cryogenic plant [5] by a vacuum jacketed distribution system [6]. At each cryomodule there are two supply and two return bayonet connections. A cryomodule has two mating bayonet connections in each end can. A typical connection in the cryomodule is a female bayonet with a ball valve for isolation and a safety relief valve. All volumes that can have cold helium trapped inside include safety relief valves. The supply end can includes a primary and shield 3.8 cm bayonets. The primary supply carries 3 bar, 2.2 K helium to a J-T valve. The J-T valve is controlled to maintain a set liquid level in the cryomodule of 2 K, 0.031 bar liquid helium. The shield supply line is a 3 atm 40 K helium line with no valves internal to the cryomodule. The return end can includes two bayonets and a primary circuit relief stack. The primary supply bayonet is a larger 7.9 cm bayonet to support pumping the required mass flow at the lower return pressure. The shield return bayonet is the same size as the supply. The relief stack is designed for high flow rates due to the large liquid inventory and potential for very high heat loads in the event of a loss of vacuum in the cryostat. There is a two stage relief setting. A lower pressure, 1.2 atm, relief is sized for the maximum mass flow through the inlet J-T valve and a higher pressure, 4.1 atm differential, relief valve is sized for the flow associated with a loss of vacuum [7]. Both relief valves are reseating designs. All cryogenic connection between air and sub-atmospheric helium include a double o-ring seal with vacuum pumping between the o-rings to ensure air is not allowed to leak into the cryogenic circuit.

### 6.2.6 Vacuum Interfaces

The CEBAF linacs are separated from the warm beam transport vacuum chambers as they include cold components that will cryopump all gasses and accumulate contamination over time. The separation of the vacuum spaces is done with a differential vacuum pump, an electrostatic precipitator, and a fast gate valve. The combination of these elements is designed to protect against off normal vacuum events in the warm beam transport chambers. The cryomodule has three separate vacuum systems, the beamline which include one 30 L per second ion pump, insulating vacuum, and the RF power coupler which includes one 20 L per second ion pump each. The cryomodule beamline interface to the warm girder is an electro-pneumatic controlled 70 mm viton sealed gate valve. The insulating vacuum has a single manually actuated 152 mm viton sealed gate valve that is connected to a pumping station for initial pump down and maintenance as needed. The RF power couplers are configured with a manifold allowing two couplers to be pumped by a single ion pump. The manifold has a 70 mm manual right angle viton o-ring sealed valve used initial pump down and maintenance operations.

### 6.2.7 Heat Load Estimates

The CEBAF cryomodule is dominated by dynamic RF heat loads. The cavity specification allows for a maximum dynamic heat load of 5.4 W per cavity. Due to improved gradient performance of the cavities over the specification the typical operating conditions for the cavities are at ~150 % of the design values [8]. Along with a dynamic waveguide heat load this accounts for the ~100 W of dynamic 2 K heat load. The static heat load budget for each cryomodule is 18 W and has been measured to be closer to 14 W on average. The static heat load has several components including the bayonets, beamline transitions from 2 to 300 K, and the RF power couplers. The shield static heat load is measured at ~150 W and comes from the heat stationing of the bayonets, beamlines transitions and the RF power couplers.

### 6.2.8 Cavity

The LE5 SRF cavity developed at Cornell's University's Newman Laboratory, has five cells resonant at 1497 MHz, one fundamental power coupler (FPC) located at one end of the cavity and two higher order mode couplers (HOM) located at the opposite end. The FPC and HOM are waveguide couplers sized for specific RF cutoff frequencies. The cavity is fabricated from high RRR sheet metal niobium. The cavity is operated submerged in a 2 K helium bath. The high RRR is needed for the improved thermal conductivity in order to minimize the temperature gradient across the sheet metal keeping the RF current carrying surface at the lowest temperature possible. The cavity was designed for high current operations and the resulting superior HOM damping performance was an important factor in selecting the design for CEBAF.

### 6.2.9 Cavity Pair

Two cavities are assembled together into a cavity pair (see Fig. 6.3) with the FPCs together at the center of the pair [9]. The cavity pair includes an inner adapter between the cavity flanges, two end adapters at the ends of the cavities, end dishes for interfacing to the helium vessel, vacuum valves, and two PFC extensions between the cavity FPCs and the helium vessel feedthrough plate. All components inside the end adapters are made from niobium. All components are joined using bolted connections with indium wire seals. The center of the cavity pair is fixed and bellows in the end dish assemblies allow for the differential thermal contraction between the niobium cavity pair and the stainless steel helium vessel.

### 6.2.10 Tuner

The SRF cavities are required to be tuned to 1497 MHz $\pm$ 50 Hz during 2 K operation. After cooldown of the cavity the frequency is required to be tuned $\pm$200 kHz to ensure operations at 1497 MHz. The frequency is controlled by a mechanical tuner (see Fig. 6.5) that can stretch or compress the cavity longitudinally changing the resonant frequency. The tuner resolution requirement is 50 Hz. With the cavity frequency sensitivity of $\sim$200 kHz per millimeter, the resulting mechanical resolution requirement is 25 nm. The tuner attaches to the first and fifth cell of the cavity and is submerged in liquid helium during operation. The tuner is driven by a stepper motor mounted external to the cryomodule that transfers torque through two rotary feedthroughs. The rotary feedthrough connects to a right angle driveshaft. The driveshaft ends in a ball screw that acts to change the length of the tuner active leg. The two cell holders are attached on one side to the active tuner leg and the opposing tuner dead leg. The tuner includes limit switches and hard stops to limit the tuner range and prevent damage to the cavity. Major disadvantages of this design include the requirement for the tuner to be submerged in the helium bath, required rotary feedthroughs on the vacuum and helium vessels, and the backlash as the tuner goes from compression to tension (in later implementation of the tuner the backlash was eliminated by biasing the tune of the cavity and operating the tuner in tension only). Early accelerated life testing was done [10] to qualify the mechanical systems.

### 6.2.11 Helium Vessel

The helium vessel has several major functions in the CM design. The first is to house the cavity pair in a bath of liquid helium. This requires the vessel to be leak tight and

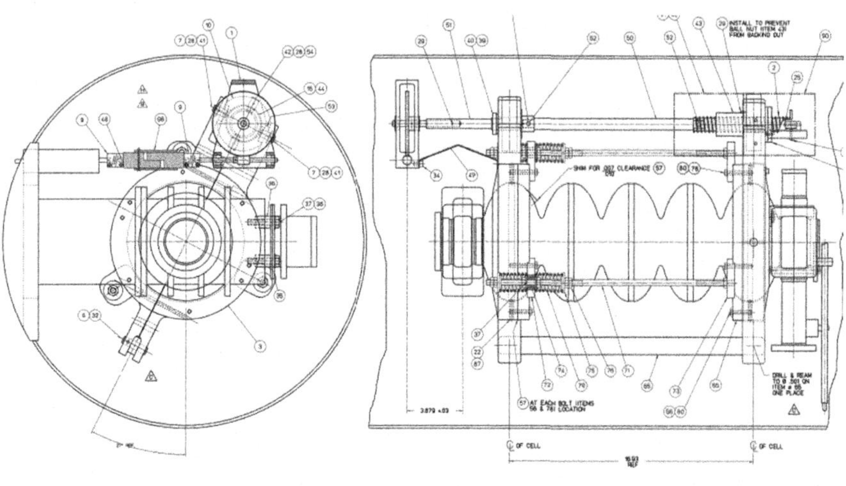

**Fig. 6.5** C20 cavity frequency tuner

operate over a range of temperature from 350 to 2 K. The second is to provide the required feedthroughs for RF power, instrumentation, and tuner mechanical drive to the cavity pair. Additionally, the helium vessel provides the support structure for the cavity pair inside the cryostat. The body of the helium vessel is a 24 in outer diameter cylinder with a massive feedthrough plate. The entire assembly is made from 316L stainless steel. The feedthrough plate is a thick plate to provide a stable interface for the waveguide indium seals, instrumentation metal gaskets seals, and welded rotary feedthrough connections. After inserting the cavity pair inside the helium vessel body the waveguide indium seals are made up and the end dish assemblies are welded to dished heads that bridge the end dishes to the helium vessel body closing the assembly. The dished head assemblies include mounting supports and a 4 in. upper and 0.75 in. lower helium connection. The mounting supports accommodate four nitronic rod on each end that spa from the vacuum vessel to the helium vessel. The large helium connection provides the cross sectional area required for heat and mass transport from one helium vessel to the next and the lower connections maintains constant liquid level throughout the four helium vessels in the CM.

### *6.2.12 Input Coupler*

The CEBAF cryomodule uses a waveguide RF fundamental Power Coupler (FPC). The FPC consists of a warm RF vacuum window that provides a hermetic seal between the RF distribution system air filled waveguide and the cryomodule vacuum waveguide, a warm waveguide assembly, a warm to cold waveguide transition [11] and a cold RF window that provides a hermetic seal between the cryomodule waveguide vacuum and the cavity vacuum. The waveguide vacuum space between the warm and cold RF windows is monitored for vacuum, arcing, and heating. The warm to cold transition waveguide assembly is a thin stainless steel rectangular waveguide with two bellows, stiffening rings for the vacuum load, a shield heat station, flanges on each end, and is copper plated on the inside. The assembly is optimized for a nominal 1500 MHz RF power of 5 kW. This determines the thickness of the copper plating and the location of the thermal intercept. As the RF power increases the optimum position for the intercept moves toward the cold end of the waveguide. The bellows are three convolutions made of thin stainless steel and located close to the ends of the waveguide. The bellows makes up for any slight misalignment between the helium vessel feedthrough plate to the sheet metal cavities. Additionally, the thin walled bellows create a thermal break in the waveguide wall. The FPC contributes 0.8 and 5.5 W to the 2 and 40 K heat loads respectively.

### *6.2.13 HOM Loads*

The LE5 cavity selection for use in CEBAF was partly due to the excellent HOM damping of the cavity and HOM loads. The HOM loads are assembled on the end of a superconducting waveguide that is mounted on the cavity pair and remains inside the helium vessel. For the CEBAF beam currents the dissipated power in a HOM absorber is a fraction of a watt making it preferable to dissipate the HOM power in the helium bath rather than bringing the heat to the shield or room temperature. The size, both cross section and length, of the waveguide is such that the higher frequency HOM RF power is transmitted to the loads while the fundamental power is cutoff and does not propagate to the HOM load. The simple design of a cutoff waveguide has many advantages for fabrication and assembly. The HOM load material is a particular challenge and was the subject of an extensive development program [12].

### 6.2.14 *Magnetic Shields—Inner and Outer*

There are two passive magnetic shields in the cryomodule. One located on the outside of the helium vessel which operates close to 2 K and a warm shield just inside the vacuum vessel that operates at room temperature. Both shields are made from the same high permeability shielding material ~0.01–0.02 in thick. Efforts are made to minimize the holes, joints, or other features that may allow magnetic field into the cavity location. Joints are made with overlaps and taped to ensure minimum leakage in the shield. Considerable effort is given to the details of the mechanical design as any cold working of the material during assembly can degrade the performance. The shield around the exterior of each helium vessel is joined with bridging material extending to the next helium vessel forming a continuous cylinder down the length of the cryomodule that is capped at each end at the end can end plate. Subsequent designs have used segment shields that have end caps after each cavity. These segmented designs provide improved shielding performance. Penetrations from the warm to cold parts of the cryomodule require holes in the shielding. The waveguide penetrations to the cavities are a particularly difficult area to shield as the waveguides themselves have a significant cross sectional area. Extensions can be used to help attenuate the field leakage around a hole in the shielding and should be used where possible. Significant effort was made to eliminate any magnetic material inside the magnetic shielding. The use of 316L stainless steel was required inside the vessel and 304 stainless steel was permitted outside of the magnetic shielding.

### 6.2.15 *Thermal Shield and Multilayer Insulation*

Multilayer insulation (MLI) is comprised of alternating layers of thin aluminized Mylar and a low thermal conductive spacer material. MLI is used between the helium vessel and thermal shield and between the thermal shield and vacuum vessel. A nominal 15–30 and 45–60 layers are used respectively. MLI is used to reduce the radiative heat load to the cryogenic circuits and to reduce the heat flux to the 2 K circuit in the case of a catastrophic loss of insulating vacuum. A copper sheet metal thermal shield is located between the 2 K and room temperature surfaces. The shield is cooled by a single straight line running the length of the cryostat. The thermal shield acts to intercept radiative heat load and is connected to all warm to cold transitions to intercept conduction heat loads.

### *6.2.16 Vacuum Vessel*

The vacuum vessel serves multiple functions in the cryomodule. The 304 stainless steel vessel is the boundary of the insulating vacuum, the support structure for the helium vessels, and provides penetrations for the RF power coupler waveguides and instrumentation feedthroughs. The vacuum vessel is a cylinder with weld flanges on both ends and a large access port aligned with the helium vessel feedthrough plate. Mounted on each weld flange are supports for the nitronic rods that connect the vacuum vessel and helium vessel as a low thermal conducting structural support. The access port is closed with a Top Hat assembly that has provisions for the RF power waveguides, cavity tuner mechanical drives, RF signals, and instrumentation feedthroughs. The Top Hat area contains almost all of the penetrations through the insulating vacuum shell.

### *6.2.17 Cryounit*

The CEBAF Cryounit (CU) is the assembly of the helium vessel, power couplers, instrumentation and vacuum vessel around a cavity pair (see Fig. 6.6). A layer of magnetic shielding and MLI are assembled onto the outside of the vessel. This is inserted into the vacuum vessel which has the thermal shield and MLI preinstalled. The helium vessel is support by four nitronic rods on each end configured in a double x pattern for alignment purposes. Two additional longitudinal nitronic restraint rods are installed to complete the installation. The nitronic rods are used to rough align the cavity flanges to the center of the vacuum vessel and tensioned to set the load on the rods for maintaining the cavity position during and after cool-down. The Vacuum vessel access port is used to install the FPC warm to cold waveguide extensions along with thermal straps and instrumentation wiring. A aluminum "Top Hat" closes the access port and has warm waveguide and window assemblies installed. The warm waveguides are connected to a vacuum manifold allowing a single ion pump to support two FPCs. This assembly maintains the hermetic seal of the cavity pair and is the building block for the cryomodules. Four CU are assembled together in a cryomodule.

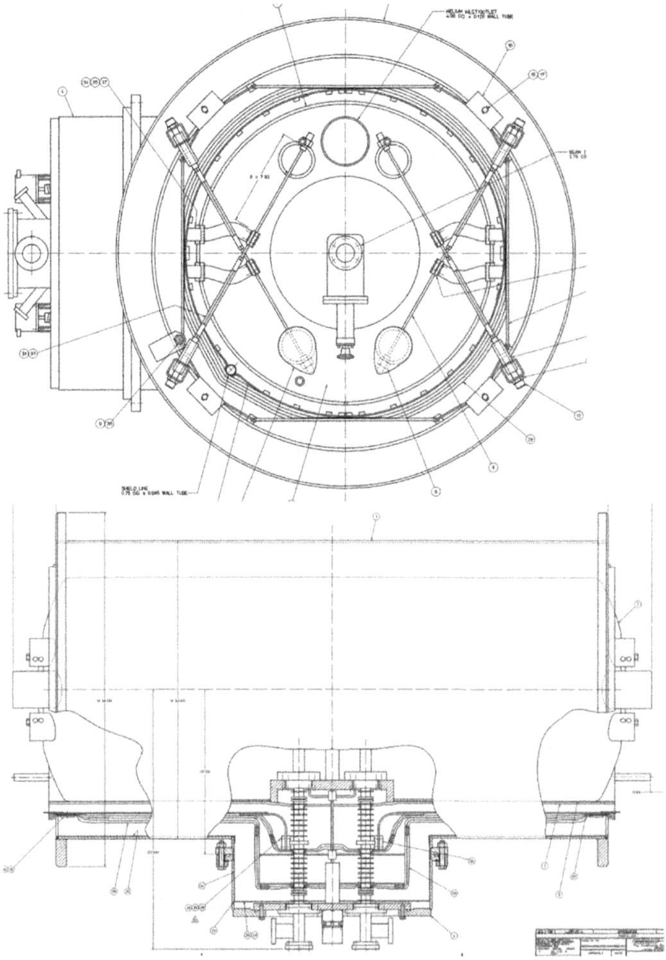

Fig. 6.6 CEBAF C20 cryounit

## 6.2.18 Instrumentation

The cryomodule cryogenic instrumentation includes temperature sensors, helium liquid level probes, helium bath pressure transducers, vacuum gauges and pumps, and a J-T actuator that is driven with a 24 VDC motor and the valve position is monitored with LVDTs. The temperature sensors are silicon diodes with $\sim 1/4$ degree accuracy. In operation the absolute accuracy of the temperature sensors has not been important as these are primarily used for controlling cooldown where rate of change is important for identifying off-normal conditions like a degraded insulating vacuum that produce large changes in temperature. The liquid level probes used are a superconducting wire type. It is important to adjust the excitation current to the operating conditions as the helium pressure changes the cooling on the superconducting wire. Warm capacitance manometer pressure transducers are used for pressure measurement. The connection of the pressure transducer to the helium bath should keep the 2 K heat load small while trying to maximize the conductance from the bath to the transducer. In the C20 design the annual space of the return bayonet is used for this connection. This adds no heat load to the system but does limit the conductance and associated bandwidth of the measurement made possible with the connection. The waveguide vacuum space is monitored for RF heating using an infrared sensor mounted inside the waveguide vacuum and isolated from the RF by a cutoff tube, RF discharge with a photomultiplier tube mounted outside the vacuum space through a transparent window mounted on a metal seal flange, and vacuum discharge using a 20 L per second ion pump. Additional vacuum monitoring is done for the insulating vacuum using a cold cathode gauge and the beamline using a 30 L per second ion pump. The waveguide instrumentation is used in the RF interlock chain while the beamline ion pump is used in the RF and beamline valve interlock chains.

### 6.2.19 Final Assembly

The cryomodule is made up from six major components, four CUs and two end cans. The CUs and end cans are assembled onto a rail where they are aligned in x and y as well as longitudinally. The location of the end cans is tightly controlled. This along with the careful control of the cryogenic distribution system allows for the use of standard size u-tubes avoiding the need for custom fabrications. The CUs and end cans are assembled with a set of "bridging" components. Each bridging section between the components has the helium circuits connected with welded connections and the MIL, thermal shield, and magnetic shields are extended, and insulating bridging ring is positioned and welded in place. The final alignment of the cavities is done by adjusting the nitronic rods which are accessed through small o-ring sealed ports located on each bridging ring.

### 6.2.20 Status

The first C20 cryomodule was installed in the CEBAF tunnel in 1990 and the machine was completed in 1993. Since that time the cryomodules have performed well above their design parameters. The original 5 pass energy of 4 GeV was exceeded and 6 GeV operations was achieved with the original cryomodules. Since that time there has been a slow degradation in gradient performance of some cryomodules. As a result a refurbishment program was initiated. The progress with cavity processing in the intervening time allowed the performance specification for cavity gradient to be increased to 12.5 MV/m. This results in a "C50" cryomodule with 50 meV energy gain per cryomodule per pass. Along with the increase in cavity performance some designs were improved during this work. The FPC waveguides had a newly designed "dogleg" section installed eliminating flashover on the ceramic RF window and the tuner mechanical coupling was modified to eliminate twisting and resulting backlash in the drive train. To date eleven cryomodules have been refurbished and the twelfth is in process.

## 6.3 The Spallation Neutron Source (SNS) Cryomodule

### *6.3.1 Introduction*

The SNS is a 1 GeV negative hydrogen ion accelerator with up to 2 MW power producing a source of neutrons for materials research. The initial portion of the acceleration is achieved via a conventional negative proton injector, a drift tube linac (DTL) and a coupled cavity linac (CCL), that provide a nominal energy gain up to 185 MeV. The machine was changed in December 1999 from a warm temperature to a cold temperature linac to improve overall machine performance. The super-conducting linac (SCL), discussed here, contains 11 medium beta cryomodules capable of 345 MeV and 12–21 high beta cryomodules capable of up to 1300 MeV. After passing through an accumulator ring the beam goes to a mercury target where a neutron beam of as much power as 2 MW is produced.

The cryomodule (CM) is based on the CEBAF CM with improvements borrowed from LHC, TESLA, and the JLab 12 GeV upgrade and uses the frequency scaled KEK fundamental power coupler (FPC). Figure 6.7 is the elevation view of the high beta CM, while Fig. 6.8 is the flow schematic. The FPC requires a 4.5 K lead flow to cool the outer conductor; therefore the LHC concept of producing the 2 K in the CM rather than in the refrigerator is utilized.

The refrigerator produces a 3 bar, 4.5 K stream, which feeds two Joule-Thomson (JT) valves in parallel. The first supplies a small sub-cooler in the CM and then cools the cavity. The second feeds the power coupler outer conductor. The CM

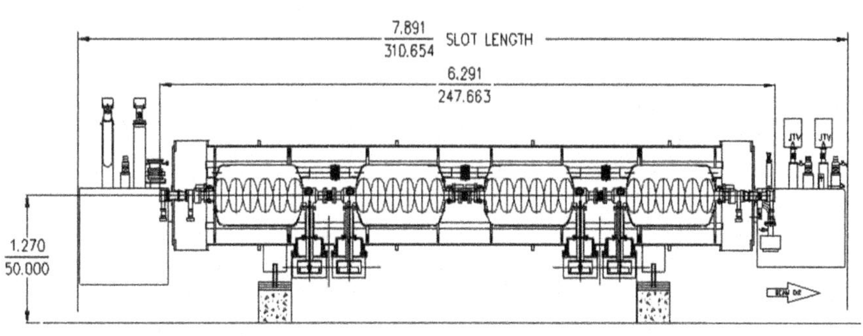


**Fig. 6.7** High beta cryomodule schematic

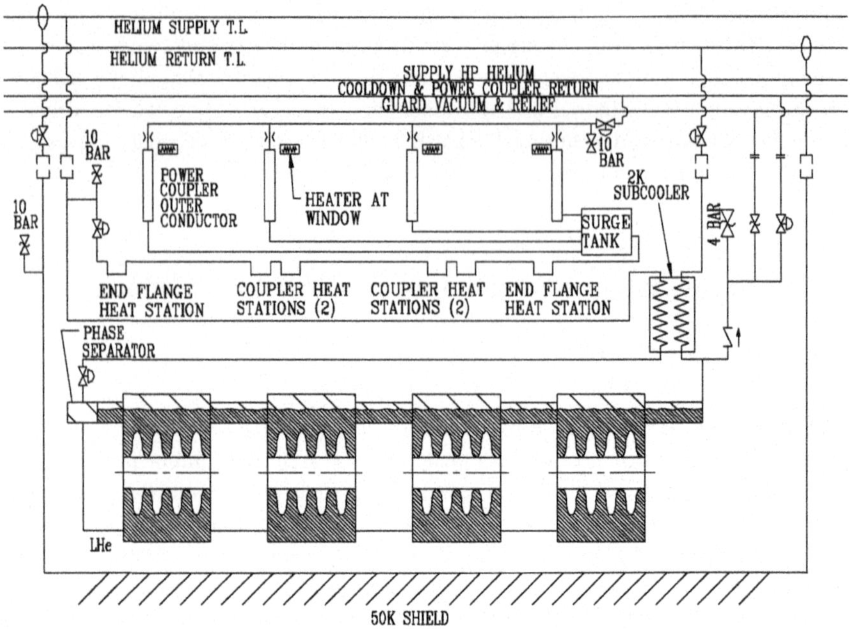


**Fig. 6.8** Flow schematic

shield is cooled by a 4 bar, 35 K stream, which first cools the supply transfer line (TL) shield, then the CM shield, and finally the return TL shield before returning to the refrigerator at 52 K. The bayonet design permits replacement of a CM in less than a day if needed without warming up the entire linac. In the nine years since the initial CEBAF cooldown, the linacs have never been warmed up and only four CMs have been replaced during scheduled accelerator shutdowns.

The relevant parameters for both medium and high beta CMs are given in Table 6.2 and the relevant refrigeration capacities are given in Table 6.3.

## 6.3.2 *Cavity String*

The design, manufacture, and performance of the SRF cavities are reported elsewhere [13]. The medium beta CM consists of three six-cell cavities while there are four six-cell cavities for the high beta CMs. The SNS cavities operate at 805 MHz. During initial cold tests, the cavities have met the design requirement for accelerating gradient, which is an $E_{\text{peak}}$ of 27.5 MV/m at a design quality factor, $Q_0$, of $6 \times 10^9$ at 2.1 K. Power dissipation per cavity at 7 % duty cycle is 2 and 3.5 W for the medium and high beta cavities respectively. The relationship of $Q_0$ with temperature follows BCS theory. Accordingly, the $Q_0$ of $14 \times 10^9$ at 2 K decreases to $6.9 \times 10^9$ at 2.3 K. The two different beta cavities are elliptical in shape,

**Table 6.2** SNS cryomodule parameters

| | Medium | High |
|---|---|---|
| Slot length | 5.839 m | 7.891 m |
| CM length (bore tube) | 4.239 m | 6.291 m |
| CM diameter | 1.22 m (~48″) | |
| 2 K Heat load (static/dynamic) | 25/14 W | 28/20 W |
| Maximum coupler flow | | 0.075 g/s |
| Shield heat load including Transfer Line | 170 W | 200 W |
| Tunnel H × W | | 10 × 14 ft |
| Control valves per CM | | 5 |
| Bayonets per CM | | 4 |
| Radiation hardness | | $10^8$ rads |
| Pressure rating | 2 K System Warm | 3 atm |
| | Cold | 5 atm |
| | Shield and 4.5 K systems | 20 atm |

**Table 6.3** Refrigeration capacities

| | He temp (K) | Capacity (W) | Pressure (atm) | Flow (g/sec) |
|---|---|---|---|---|
| Linac shields | 35–52 | 8300 | 4.0 | 90 |
| Linac cavities | 2.1 | 2400 | 0.041 | 120 |
| Secondary | 4.5 | | 3.0 | 0.15 |

manufactured from 4 mm thick niobium and have stiffening rings at 80 mm to minimize microphonics. The performance of the first prototype medium beta cavity, shown in Fig. 6.9, exceeded the required performance.

The cavities are housed in a titanium helium vessel, which matches the coefficient of thermal expansion of the cavity. The cavity is maintained at operating frequency through a TESLA-style tuner mounted on one end of the helium vessel, which operates through a bellows. The tuner, manufactured out of stainless steel, is actuated through a cold stepping motor through a harmonic drive. Power is brought into the cavity through the coaxial FPC [14] at one end of the cavity. Higher order mode extraction filters are attached at each end of the cavity to damp some potentially dangerous longitudinal modes. The string of cavities in the helium vessel with couplers, HOM damping filters and hermetic valves are assembled in JLab Class 100 clean room to minimize contamination. Figure 6.10 shows the arrangement of the cavity string including the cavity, coupler, helium vessel and tuner.

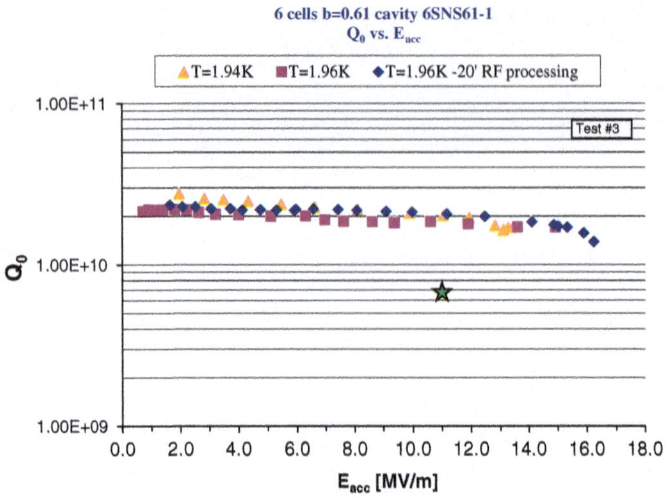


**Fig. 6.9** Medium beta cavity performance

**Fig. 6.10** Cavity string components inside the space frame situated on tooling

## 6.3.3 *Cryomodule*

A high beta CM consists of two end cans, a vacuum tank, a space frame, a thermal shield, two magnetic shields and a hermetically sealed string of four cavities in the helium vessel each with FPCs, a field probe, two HOM filters, bellows between cavities and seal valves. A medium beta CM is similar in construction but houses a string of three hermetically sealed cavities instead of four. The general arrangement of these components is shown in Figs. 6.11 and 6.12. The CMs used in CEBAF, the

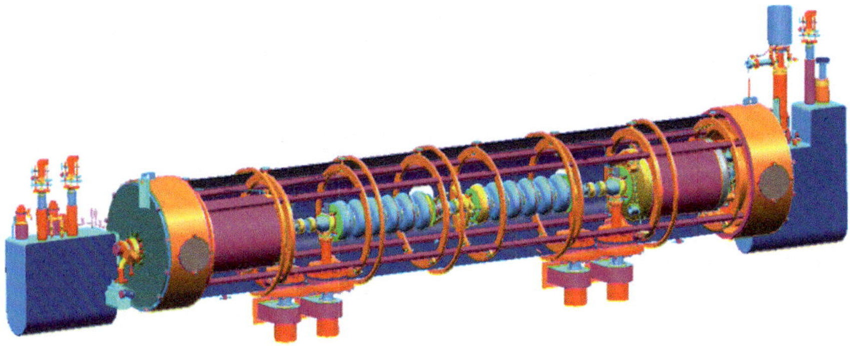

**Fig. 6.11** High beta cut-away

**Fig. 6.12** Medium beta cut-away

largest use of SRF in the US, employs a similar construction arrangement. What makes the SNS design unique is that the CMs are assembled at JLab and shipped to the SNS site at Oak Ridge National Laboratory (ORNL). This led to the development of a structure that can handle the over-the-road shipment of 500 miles while maintaining alignment of the cavities in a low loss cryostat [15]. The space frame was developed at JLab for the next generation JLab CM to facilitate installation of long strings of cavities efficiently at a relatively low cost. The space frame was adapted and strengthened to handle the relatively high transportation loads incurred during shipment to ORNL. The vacuum tank provides a support structure for the space frame and minimizes the gaseous conduction to the cold surfaces on the cavity string. The string is supported off of the space frame through nitronic rods, which limit the solid conduction between room temperature and the cryogenic surfaces. These rods are also able to take the shipping loads and maintain the cavities in alignment. A thermal shield, operating at 50 K, surrounded with multilayer insulation (MLI) provides a radiation barrier between the cavities and the outside world. There are two magnetic shields, one outside the space frame and the other at the helium vessel, which in concert reduce the earth's and stray magnetic fields by a factor of ~100 to minimize the effect on the cavity $Q_0$. L shaped end cans, a design developed for CEBAF to save space, close off the cavity string in the vacuum tank and provides the interface for the helium to cool the cavities, the couplers and the thermal shields. Between each CM is a 1.6 m warm space that contains quadrupole magnets and diagnostics including beam position monitors, current transformers and wire scanners.

## 6.3.4 Cryomodule Heat Loads and Thermal Design

The design parameters for the CM were developed in conjunction with the design parameters for both the cryogenic system and the SC portion of the linac and are given in Table 6.2. Specifically, the heat load budgets were developed considering operational experience with the CEBAF cryogenic system, where applicable. The heat load estimates and the budget for the CM are given in Table 6.4. This section contains a summary of the heat loads as well as detailed descriptions of the contributions of each subsystem.

### 6.3.4.1 Cavity, Helium Vessel and Tuner

The SNS helium vessel design, described in [16], supports the cavity during all phases of operation, facilitates cavity tuning and contains the cryogens with appropriate plumbing. Each 0.61 m diameter titanium helium vessel contains a single niobium cavity whose six cells are immersed in ~150 L of He-II during normal operation. The ends of the cavities protrude through the helium vessel heads into the insulating vacuum to allow beamline assembly as well as attachment of FPCs, Higher Order Mode (HOM) filters and field probes. These end groups are subject to radiation heat transfer from the 50 K shield and the FPC, solid heat conduction from instrumentation leads and the FPC as well as power generated within the HOMs. Extraction of HOM signals to room temperature is planned, requiring a 3.6 mm semi-rigid coaxial cable running from the cavity to the vacuum tank exterior. Based on finite element analysis, the heat loads incident on the HOMs must not exceed 0.25 W in order to maintain the HOMs well below the niobium critical temperature. Additionally, instrument wiring for temperature diodes, liquid level sensors, and heaters are routed from the helium vessel to the vacuum tank exterior. The helium vessel heads possess attachment points for Nitronic-50™ stainless steel support rods and the cavity tuner. The calculated heat load intercepted by the cavity and helium vessel assembly is given in Table 6.4.

The dynamic heat load for each cavity design has been calculated for $E_{peak}$ = 27.5 MV/m. It is desirable to increase the accelerating gradient such that $E_{peak}$ = 35 MV/m, thereby reducing the number of high beta cryomodules required

**Table 6.4** Calculated heat loads (W) for the medium and high β cryomodules

| Estimate summary | | MBCM | | | HBCM | |
|---|---|---|---|---|---|---|
| | Qty | 2.1 K | 50 K | Qty | 2.1 K | 50 K |
| Static—(see itemized list) | 1 | 9.7 | 131 | 1 | 11.5 | 161 |
| U-Tube Allotment | 1 | 10.0 | 24 | 1 | 10.0 | 30 |
| Dynamic (see itemized list) | 1 | 8.3 | | 1 | 17.1 | |
| Total heat load per CM | | 28.0 | 155 | | 38.6 | 191 |
| Budget per CM | | 39.0 | 170 | | 48.0 | 200 |
| *Dynamic contributions* | | | | | | |
| Cavity | 3 | 6.0 | | 4 | 14.0 | |
| Power coupler | 3 | 0.6 | | 4 | 0.8 | |
| Bellows | 2 | 0.2 | | 3 | 0.3 | |
| HOM (2 per cavity) | 3 | 1.5 | | 4 | 2 | |
| Total dynamic | | 8.3 | | | 17.1 | |
| *Static contributions* | | | | | | |
| Radiation—HV and bellows | 3 | 1.1 | 41.7 | 4 | 1.8 | 65.3 |
| Power coupler (radiation) | 3 | 2.1 | | 4 | 2.8 | |
| Tuner | 3 | 0.75 | | 4 | 1 | |
| He vessel supports | 3 | 0.2 | 18 | 4 | 0.3 | 24 |
| Warm beam tube conduction | 2 | 0.1 | 2 | 2 | 0.1 | 2.5 |
| Warm beam tube radiation | 2 | 0.9 | 0.9 | 2 | 0.9 | 0.9 |
| Cables (3 per cavity) | 1 | 0.5 | 1.8 | 1 | 0.5 | 1.8 |
| Supply bayonets | 2 | 1.0 | 12 | 2 | 1.0 | 12 |
| Radiation | 1 | 0.04 | 9.0 | 1 | 0.04 | 9.0 |
| PC J-T valve | 1 | 0.25 | 2 | 1 | 0.25 | 2 |
| Subcooler J-T valve | 1 | 0.25 | 2 | 1 | 0.25 | 2 |
| Shield relief | 2 | 0.0 | 4 | 2 | 0.0 | 4 |
| 5 K transfer line | 1 | 0.1 | | 1 | 0.1 | |
| 50 K transfer line | 1 | | 3 | 1 | | 3 |
| Return bayonets | 2 | 1.5 | 6 | 2 | 1.5 | 6 |
| Radiation | 1 | 0.1 | 11.2 | 1 | 0.1 | 11.2 |
| Cooldown/PC return | 1 | 0.25 | 10 | 1 | 0.25 | 10 |
| Shield relief | 1 | 0.3 | 2 | 1 | 0.3 | 2 |
| Cooldown valve | 1 | 0.25 | 2 | 1 | 0.25 | 2 |
| 5 K transfer line | 1 | 0.0 | | 1 | 0.0 | |
| 50 K transfer line | 1 | | 3 | 1 | | 3 |
| Total static | | 9.7 | 131 | | 11.5 | 161 |

to reach a given linac output energy. The dynamic heat load increases at least quadratically with increasing peak gradients or higher if operated in the field emission region. These increased losses would be handled by some portion of the available margin in the primary circuit of the refrigeration system.

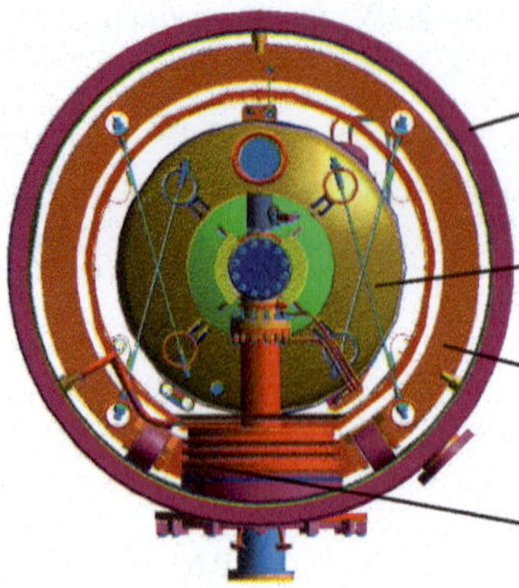


**Fig. 6.13** Nitronic-50™ support rods attach the helium vessel assembly to the space frame

#### 6.3.4.2 Vacuum Vessel and Space Frame

The vacuum vessel and space frame, described in [17], serve to locate the cavities accurately within the CM via the Nitronic-50™ support rods, provide the structural links to the external supports and remain at room temperature (300 K) during normal operation (Fig. 6.13). The 0.99 m diameter vacuum vessel contains the insulating vacuum and provides pressure containment in the unlikely event of cryogenic piping failure. The insulating vacuum is cryo-pumped during accelerator operation to less than $1.3 \times 10^{-9}$ atm, an adequate vacuum level to minimize residual gas conduction. If a cryogenic piping failure occurs, two spring-loaded parallel plate pressure reliefs, identical to those used in the CEBAF CM, open at 1.2 atm on the vacuum vessel. In addition, there are many flanged penetrations in the vacuum vessel: ports in the bottom of the vacuum vessel through which the FPC provides RF power to the cavities, instrumentation ports containing electrical feed-throughs for diodes, cavity heaters, tuner power and limit switches, field probes, and connections for the warm helium coupler exhaust.

#### 6.3.4.3 Thermal Radiation Shield

As in CEBAF, a copper shield, 2.37 mm thick, operating between 35 and 50 K is used to intercept thermal radiation (Fig. 6.14). The shield is cooled by supercritical helium gas (4 atm, 35 K) through a single pipe, 22.2 mm outer diameter by 1 mm thick wall. Thermal analysis of both single-pass and three-pass shield cooling showed $\sim$2 % increase in the total static heat load to 2 K with a single-pass cooling scheme. Cost ultimately dictated the choice of single pass versus three-pass piping.

The shield is divided into segments that cover the helium vessel assemblies and bridges that cover both the cavity-cavity and cavity-end-can interconnect regions. The bridge sections allow access to the cavity string for assembly, alignment, instrumentation wire routing and tuner maintenance. There are three segments in a

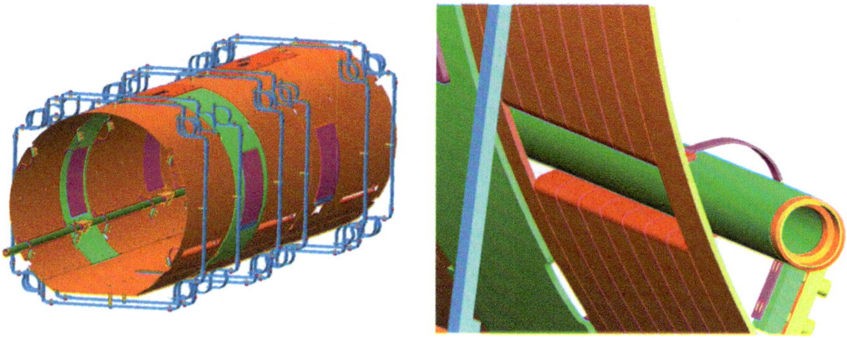

**Fig. 6.14** MBCM thermal shield assembly (*left*) and piping strain-relief (*right*)

medium beta cryomodule and four segments in a high beta cryomodule. A hydro-formed stainless steel bellows, soldered with EasyFlo-45™, connects the copper piping between segments. In addition to the bellows, a series of tabs 76 mm long × 76 mm wide, where the pipe is soldered with BCuP to the shield, provides strain relief from differential thermal contraction of the piping and shield during cool down and operation.

The horizontal and vertical helium vessel Nitronic-50™ support rods are heat-stationed to the shield segments by copper straps, 1 mm × 25 mm × 125 mm long. All wiring and cabling routed to 2 K is heat stationed at 50 K on the nearest shield segment as well. Due to the long length (>600 mm), small cross-sectional area (20 $mm^2$) and difficulty during assembly, the two axial support rods are not heat stationed. The additional heat load to 2 K is negligible (<1 mW/support).

Each bridge is fixed at only one end to an adjoining segment with PEM fasteners and allowed to slide over the other neighboring segment. Each segment-bridge assembly weighing ~45 kg is supported by eight G10 straps, 38 mm × 4.8 mm × 508 mm long, terminated with stainless steel pipe clamps affixed to the space frame support tubes. The four horizontally oriented supports react lateral shipping and handling loads, the four vertically oriented supports react both gravity and transportation loads and all eight react any axial loads.

#### 6.3.4.4 Multilayer Insulation (MLI)

The insulation system design is identical in principle to that employed in the original CEBAF CM and is expected to achieve the same thermal performance. The insulation scheme reduces the radiative heat load to the cold surfaces, provides ample mass and heat capacity to mitigate thermal transients and is comprised of materials suitable for use in a high radiation environment.

Blankets are constructed of alternating layers of double-aluminized Mylar™ (DAM), 25 μm thick, and Reemay™ #2250, a spun-bonded polyester 75 μm thick. The maximum emissivity allowed by emissometer measurement is 0.035 at room temperature. The surfaces of the helium vessel are covered with two 12-layer blankets, the beamline components and piping are spiral-wrapped with 15 layers at a 50 % overlap, and the 50 K surfaces are covered with four 15-layer blankets. The joints and seams, staggered both axially and around the circumference by 25–50 mm, are closed with 25.4 mm wide aluminized Mylar™ tape.

Effective thermal conductance through the blanket layers is calculated considering three mechanisms: radiative heat transfer—governed by emissivity (ε) and geometry, solid conduction—governed by effective thermal conductivity ($k_{eff}$) and geometry, and residual gas conduction—governed by average local pressure. A Fortran program, TRANSAM, written originally for transfer line design was used to estimate the expected heat flux to the 50 and 2 K surfaces. Using a worst-case scenario with average pressure of $10^{-4}$ torr, $\varepsilon = 0.2$ and $k_{eff} = 1.5$ mW/cm K, the heat flux to the 50 K thermal shield was calculated conservatively as 2.5 W/m$^2$. Using the same average pressure, $\varepsilon = 0.06$ (lowered since emissivity typically decreases with decreasing temperature) and $k_{eff} = 1.5$ mW/cm K, the heat flux to the 2 K surfaces was calculated to be 94 mW/m$^2$. While these values are conservative, proper assembly and installation of the blankets is required for ideal performance. Taped joints, insufficient overlap and compaction of blanket layers can reduce the effectiveness of the MLI thereby increasing significantly the incident heat load. Combining the realities of assembly with the fact that nine of twenty-two measured CEBAF CMs, as reported in [18], exceeded the 50 K static heat load design criteria, it was decided to increase the estimated radiation heat load to 200 % of the calculated values.

#### 6.3.4.5 End Cans, Heat Exchanger and Cryogenic Piping

The end cans route helium to and from the CM, provide controls for the primary and secondary circuits, provide beamline vacuum connections and contain pressure relief valves used in the event of loss of vacuum (Figs. 6.11 and 6.12). The end cans use bayonet connections that are identical to those used in the CEBAF CM [19]. In addition to the primary and shield supply bayonet connections, two J-T valves similar to those used in CEBAF are installed in the supply end can: the primary J-T has a $C_v = 0.3$ and the secondary J-T has a $C_v = 0.05$. The return end can contains a sub-atmospheric primary return bayonet, the shield return bayonet, a Circle Seal™ pressure relief valve plumbed in parallel with a parallel plate relief valve, a cool-down valve with a $C_v = 3.0$ and a helium-helium heat exchanger (HX).

The counter flow HX (Fig. 6.15) is a plate-fin type core constructed of aluminum with a pressure rating of 12 atm. Stainless-to-aluminum joints transition from the aluminum body to the stainless steel process piping. The design parameters are given in Table 6.5. Silicon diodes are installed in the assembly to measure HX terminal temperatures in order to assess the HX effectiveness.

The stainless piping within the CM is strain-relieved from the end can connections with flexible metal hose. All of the shield circuits, the fill-lines between

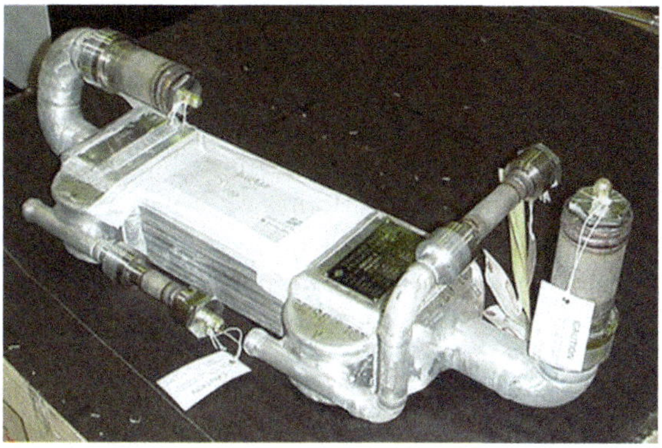

**Fig. 6.15** Helium-helium heat exchanger

**Table 6.5** Heat exchanger design parameters

| | High pressure side | Low pressure side |
|---|---|---|
| Core length | 600 mm | – |
| Core cross section | 100 mm × 100 mm | – |
| Maximum allowable helium leak rate | $1 \times 10^{-9}$ mbar l/s external<br>$1 \times 10^{-4}$ mbar l/s cross-passage | – |
| Pipe size | 26.7 mm OD × 2.1 mm wall (¾″ IPS Schedule 10) | 60.3 mm OD × 2.8 mm wall (2″ IPS Schedule 10) |
| Maximum flow rate | 6.0 g/s | 6.0 g/s |
| Fluid inlet pressure | 2.95 atm | 0.040 atm |
| Maximum pressure drop | 0.020 atm | 0.001 atm |
| Inlet temperature | 5.0 K | 2.1 K |
| Outlet temperature | 2.2 K | 3.96 K |
| Capacity | 60 W | – |
| NTU (integrated) | 4.37 | – |
| Effectiveness | 0.97 | – |
| UA (integrated) and UA margin | 93.2 W/K and 20 % | – |

helium vessels and the primary circuit plumbing through the HX high-pressure side are 26.7 mm OD × 2.1 mm wall. The primary return piping through the HX and in the relief stack is 60.3 mm OD × 2.7 mm wall, sized for a catastrophic loss of beamline vacuum. The helium vessel return headers are 88.9 mm OD × 3.0 mm wall, and along with the helium vessel provide 10 % ullage. The secondary circuit consists of 6.35 mm outer diameter × 1.3 mm wall stainless steel tubing, sized to reduce helium inventory in that circuit, and a 1 L surge tank which functions to damp potential flow oscillations that may arise during operation.

#### 6.3.4.6 Fundamental Power Coupler

Due to operating frequency and space constraints, a coaxial coupler with inner and outer conductors was selected instead of a waveguide coupler to provide fundamental RF power to the cavities. Besides delivering RF power, the coupler must not adversely affect the electromagnetic performance of the cavity or the thermal performance of the cryomodule. Cavity cleanliness procedures require the coupler to be inserted into the cavity at the six o'clock position.

To handle static and dynamic heat loads in the coupler and cavity, the outer conductor is cooled by a nominal stream of 3 atm, 5 K supercritical helium flowing at 0.038 g/s per coupler and the inner conductor is conduction cooled by a 1 L per minute water flow. The ceramic RF window is maintained at 300 K during operation by a heater-thermocouple control loop.

The outer conductor is a machined and welded stainless steel assembly with flanged connections to the cavity and the warm window. The internal vacuum surface is copper-plated with a nominal thickness of 15 μm and an estimated RRR = 10 to reduce resistive wall losses induced by the RF surface currents. A single helical, square-grooved flow passage, 2.3 mm wide × 1.6 mm deep (Fig. 6.14), with a pitch of 2.6 turns/cm is machined into a thick-walled stainless steel tube, 75.75 mm internal diameter. A thin-walled stainless steel tube (1.6 mm O.D.) is then shrunk-fit over the outer diameter of the flow passages and welded leak tight onto the ends nearest the flanges. Due to the small hydraulic diameter of 2.16 mm, flow velocities and Reynolds number are kept high to mitigate potential deleterious effects of buoyancy, re-circulation and poor heat transfer in the helium stream.

Conduction heat load estimates (Table 6.3) were calculated for the outer conductor using an Excel model that included combined conduction and convection heat transfer as well as temperature-dependent thermal properties for stainless, copper and supercritical helium gas (Fig. 6.16). A separate estimate of the static radiation heat load emitted from the FPC to the beam pipe was calculated.

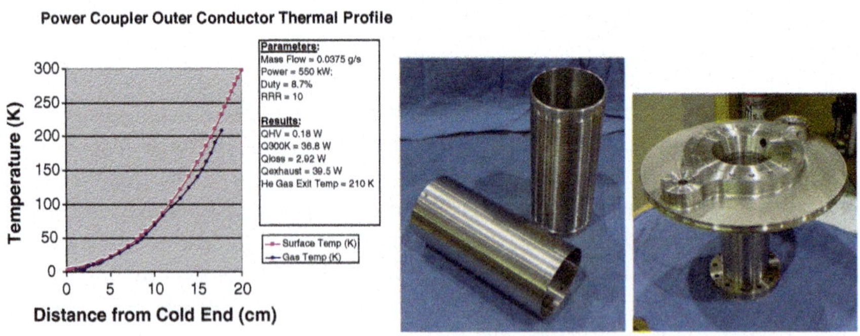


**Fig. 6.16** He gas cooling passages (*right*) and FPC outer conductor assembly (*left*). The helium gas flow rate of 0.038 g/s produces an exhaust temperature of 210 K (*center*)

The values were used as input to a finite element model of the cavity ends to verify that they remain below the niobium critical temperature.

In an effort to determine if thermo-acoustic oscillations or flow instabilities exist in the secondary circuit, a test rig was been designed and built.

## 6.3.5 *Thermal Performance of the SNS Cryomodule*

### 6.3.5.1 Budgeted Versus Estimated Heat Loads

For the primary circuit maintained at 2.1 K, the heat loads are divided into three categories: (1) a static contribution consisting of solid conduction, residual gas conduction and radiation, (2) a dynamic contribution consisting of RF heating in the cavity and coupler and (3) a distribution system allotment that consists of u-tube and transfer line heat loads. There is no dynamic contribution to the 50-K shield and its cooling circuit. Details of the estimated heat loads are given in Table 6.3. Budgeted heat loads in contrast, are those that each component are allowed to have and are used in the sizing of the cryogenic refrigeration system. A comparison of the budgeted and estimated heat loads is provided in Table 6.6. For the medium beta cryomodule, the total estimated heat load to the primary circuit is 72 % of the budget per cryomodule, and for the shield circuit is 91 % of the budget per cryomodule. The primary static heat load for the medium beta cryomodule is estimated as 9.7 W and includes contributions from bayonets, valves, supports, wiring, cabling and radiation. The dynamic heat load is estimated as 8.3 W and is divided evenly between the three cavities and FPCs housed within the cryostat.

### 6.3.5.2 Measured 2 K Static and RF Heat Loads

The following measurement technique, similar to that employed during CEBAF CM acceptance testing [5], is used to quantify the static heat load and RF heat load. With the primary J-T supply valve and u-tube return valve closed, the

**Table 6.6** Comparison of heat load budget versus estimate for both SNS cryomodules

| | β = 0.61 cryomodule (3 cavities) | | | | β = 0.81 cryomodule (4 cavities) | | | |
|---|---|---|---|---|---|---|---|---|
| | 2.1 K | | 50 K | | 2.1 K | | 50 K | |
| Description | Budget | Estimate | Budget | Estimate | Budget | Estimate | Budget | Estimate |
| Static | 15 | 9.7 | 146 | 131 | 18 | 11.5 | 170 | 161 |
| Dynamic | 16 | 8.3 | N/A | N/A | 28 | 17.1 | N/A | N/A |
| U-Tubes and distribution | 10 | <10 | 24 | <24 | 10 | <10 | 30 | <30 |
| Total | 41 | <28 | 170 | <155 | 66 | 38.6 | 200 | <191 |

helium bath pressure rate-of-rise is measured during three conditions: with the helium bath heater on (HEATER), with only RF power on (RF) and with no heater or RF power (STATIC). Five separate 30-second measurements are made in the following order—STATIC, HEATER, STATIC, RF and then STATIC. Using a known heater power ($Q_{heater}$) and the average rate-of-rise (dP/dt) for each condition, the static and RF heat loads are then calculated from the following equations:

$$Q_{static} = Q_{heater} * \left(dP/dt|_{static}\right) / \left(dP/dt|_{heater+static} - dP/dt|_{static}\right) \tag{6.1}$$

$$Q_{RF} = Q_{heater} * \left(dP/dt|_{RF+static} - dP/dt|_{static}\right) / \left(dP/dt|_{heater+static} - dP/dt|_{static}\right). \tag{6.2}$$

Note that during this measurement, the flow to the secondary circuit for the FPC outer conductors ideally should be maintained constant. Actually the inlet temperature to the CM increases during the two-and-one-half-minute data acquisition cycle. Typically several measurements are taken before opening valves to recover the pressure. Over that time span the inlet temperature can increase by as much as 5–7 K. There is a corresponding rise in the FPC cold flange temperature. This transient temperature more closely reflects the fluid temperature rather than the cold flange temperature since measurements have been made at relatively high gradients without indication of quenches. The helium bath pressure initially is approximately 3 kPa (0.030 atm) and may rise to as high as 4.1 kPa (0.040 atm) during the series of measurements. The actual total pressure rise is primarily dependent on the cavity gradients selected for each measurement.

The measurements are accurate to ±½ W due in part to the accuracy of the capacitance manometer pressure transducer as well as the measured voltage and current on the cavity bath heater. Effects from valve leakage and reduced flow in the supply bayonet have been neglected, which result in conservative heat load estimates. Overall, the static heat load measurements agree well with the estimates; the average static load is approximately 10 W (Fig. 6.17).

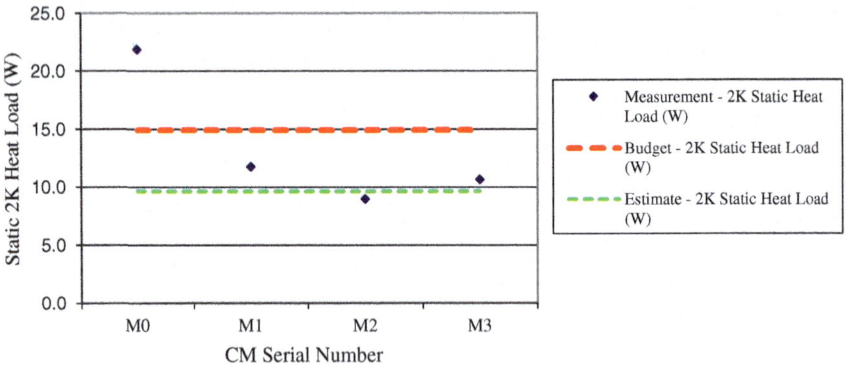


**Fig. 6.17** Measured primary static heat load versus CM serial number

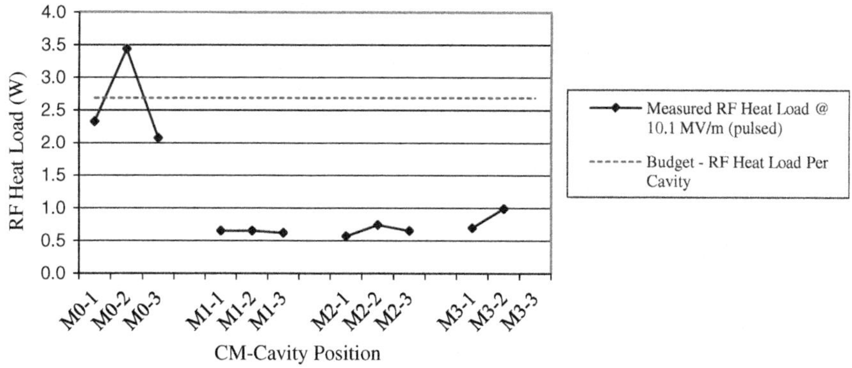


**Fig. 6.18** Measured RF heat load versus CM-cavity position at 10.1 MV/m

The measured RF heat load at the required gradient of 10.1 MV/m is approximately equivalent to the estimate for the prototype cavities, 2.7 W, and is less than 1 W for each of the cavities in the three production CMs (Fig. 6.18). One cavity, M3-3, was not measured due to schedule constraints.

The power dissipation scales quadratically with accelerating gradient and is inversely proportional to a geometric factor. For the $\beta = 0.81$ cavities, the required accelerating gradient is 15.5 MV/m and the geometric factor is reduced by 12 %. Assuming no degradation due to field emission, the power dissipation will be approximately two times greater for the higher beta cavities. Initial vertical tests on these cavities have yielded acceptable results.

#### 6.3.5.3 Measured 50 K Shield Heat Load

The measurements of the 50-K shield heat load are in good agreement with predictions (Fig. 6.19). The measurement procedure included the following steps: first for a given valve position, monitor the gaseous helium inlet and outlet temperatures and the mass flow rate, then calculate the heat load using gaseous helium properties and time-averaged temperatures and flow rates. This procedure was repeated at several different flow rates. The accuracy of this type of measurement, approximately $\pm 10$ %, is driven by the helium mass flow meter.

The heat load measured in the prototype CM (M0) is 129 % higher compared with measurements on the production CMs. This is most likely caused by increased residual gas conduction in the insulating vacuum space due to the large helium leak rate that had been detected in the prototype shield process piping after cooldown. The prototype required active turbo-pumping during the entire test period to maintain the vacuum in the range of $10^{-6}$ torr. In addition, changes in the multilayer insulation blanket design and improved blanket installation techniques contributed to the lower overall heat load in the production CMs.

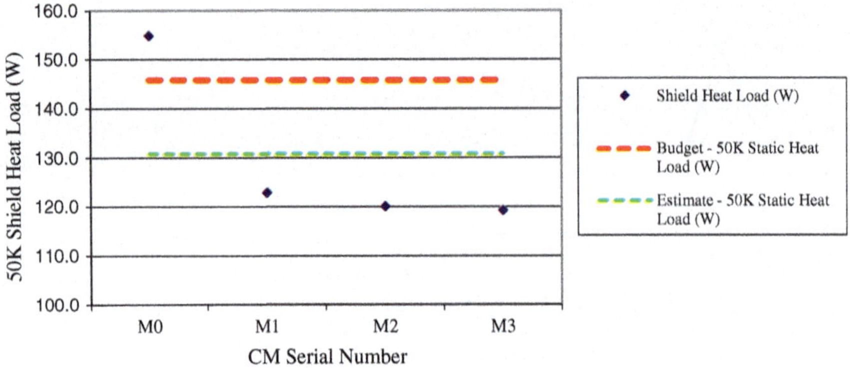


**Fig. 6.19** Measured shield heat load versus CM serial number

#### 6.3.5.4 Fundamental Power Coupler Thermal Performance

The primary requirement for the FPC is to transmit a maximum of 550 kW of RF power at a 7 % duty cycle. Adding a 10 % engineering margin, the design goal for the average power handling capability became 53 kW. Another important requirement is to insert the coupler from the bottom to minimize particulate introduction and generation in the cavities. The outer conductor has counterflowing helium-gas cooling to intercept the static and RF heat loads. The maximum flow rate is 0.075 g per second per coupler. The flow passages are sized to keep the flow velocity high enough to overcome buoyancy forces and potential flow instabilities.

To verify the design and confirm the thermal calculations, the FPC installed on the M0-2 cavity was instrumented to measure the temperature profile along the outer conductor. In addition, the exhaust helium mass flow rate was measured during this particular test to determine the maximum power handling capability of the FPC. With a stub tuner installed, the FPC transmitted 9 kW of CW power to the cavity. The resonant condition created by the stub tuner settings resulted in a transmitted power equivalent to 100 kW within the FPC, nearly double the design goal.

The thermal model used the cold and warm end temperatures, the helium mass flow rate and the equivalent transmitted RF power during the test as inputs. The results from the thermal model agree well with the experimental temperatures (Fig. 6.20). The calculated conductive heat load transmitted to the cavity from the outer conductor during the higher power operation was estimated as 1.1 W—the expected conductive heat load at nominal operating conditions is less than 0.25 W.

#### 6.3.5.5 Helium Heat Exchanger Performance

The HX is located in the CM return end can to increase overall cryogenic system efficiency. This enables the SNS supply transfer lines to contain 5-K supercritical

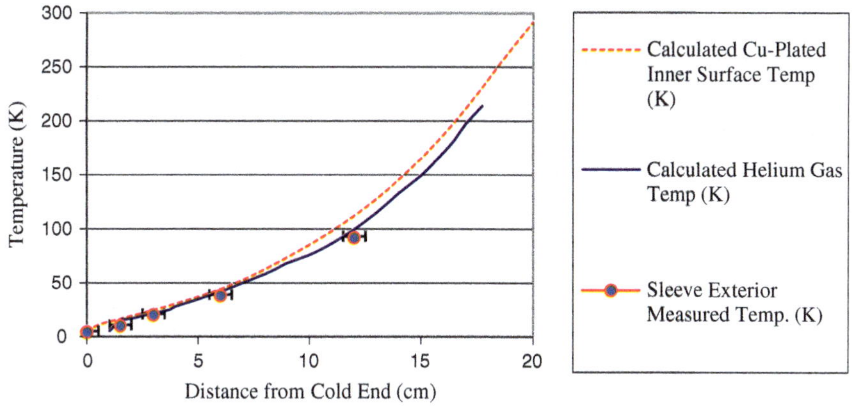


**Fig. 6.20** Comparison of analytical and measured temperatures on the M0-2 FPC outer conductor

helium as compared with ~3 K for the CEBAF machine. The HX, a brazed-aluminum plate-fin construction, pre-cools the helium upstream of primary JT valve to increase liquid yield. The effectiveness is specified as greater than 90 % for the design capacity of 60 W. The overall thermal conductance (UA) and number of transfer units (NTU) are 93.2 W/K and 4.37, respectively. The assembly is required to be leak tight to $1 \times 10^{-9}$ mbar l/s.

Initial measurements indicate that the performance is consistent with estimates. The measured inlet and outlet temperatures on the low-pressure-side of the HX are 2.0 and 3.5 K respectively, compared with estimates of 2.1 and 3.96 K. Inlet and outlet temperatures for both streams are required to calculate the effectiveness of the counter-flow HX. Since the high-pressure-side inlet temperature indication is not reliable due to the placement of the silicon-diode temperature sensor, the effectiveness has not been calculated.

#### 6.3.5.6 Conclusions on the Thermal Performance

Measurements of thermal performance have been completed on the prototype and three production CMs. Results obtained for the static, RF and shield heat loads are in good agreement with estimates and are within budgeted heat loads. Measurements of the outer conductor thermal profile are in good agreement with calculations and indicate that the heat load from the FPC is acceptable. The FPC and its outer conductor can transmit approximately two times more than the required maximum RF power expected from the SNS klystrons. Initial measurements of the available heat exchanger terminal temperatures agree with predictions. The heat exchanger performance assessment is incomplete; more study is required to quantify the heat exchanger effectiveness.

In addition, better CM and cavity construction techniques and procedures have incrementally improved the thermal performance.

## 6.4 The CEBAF C100 Energy Upgrade Cryomodule

### *6.4.1 Introduction*

In 2008 construction began on the energy upgrade of the CEBAF accelerator to 12 GeV [20]. To achieve the 12 GeV energy five new cryomodules were added to each of the two existing linacs. Each cryomodule provides 100 MV of accelerating voltage with eight 1500 MHz SRF cavities. Each cavity contains seven 1500 MHz cells, two hook style HOM couplers, and a FPC [21]. Each cavity has 0.7 m of active length and is required to operate at 19.2 MV/m to provide the 100 MV of required voltage and have sufficient voltage overhead for robust operations. A new low loss cell geometry [22] is used to reduce the 2 K heat load. The cavities, cryomodule, and helium refrigerator are designed for a total of 300 W 2 K heat load for each cryomodule. The cavity and cryomodule alignment requirements remain unchanged from the original CEBAF specification [23]. The C100 cryomodules (see Fig. 6.21) are installed in existing empty linac zones and are required to maintain the same cryogenic circuits and interfaces as the C20 cryomodules (see Fig. 6.22).

### *6.4.2 Lessons Learned from C20 Experience*

The C100 cryomodule design benefited from the production and decade of operations of the C20 cryomodules as well as advances in the SRF technology and

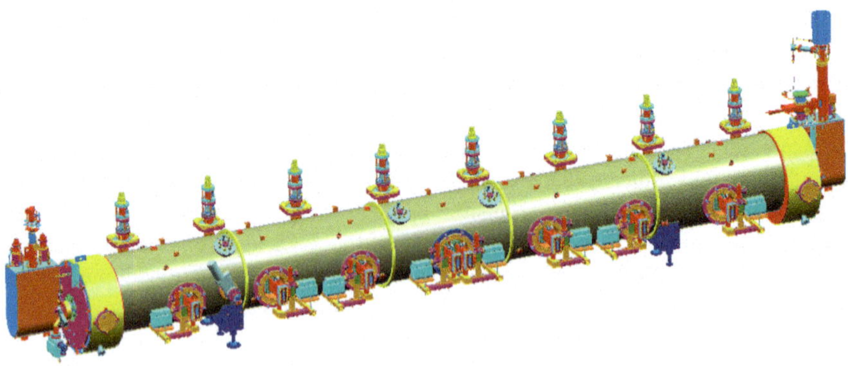

**Fig. 6.21** C100 cryomodule assembly

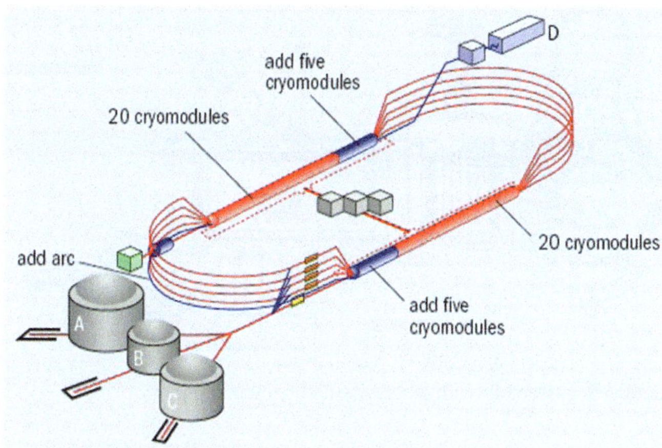


**Fig. 6.22** 12 GeV CEBAF with ten C100 cryomodules

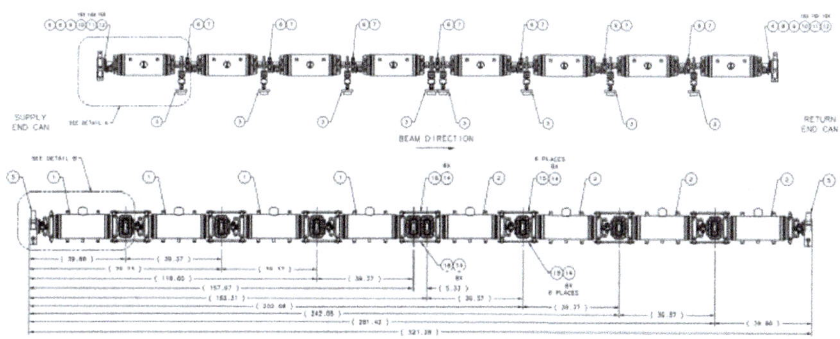

**Fig. 6.23** C100 cavity string

facilities. The largest change in the C100 cryomodule was the move away from a cavity pair assembly used as the basis for the C20 cryomodule to an eight cavity string (see Fig. 6.23) assembled in the clean room. The cavity string assembly is too large to be tested prior to assembly into a cryomodule. This requires each cavity to be vented after being tested individually prior to being assembled into the cavity string. This use of a cavity string assembly is based on a high level of confidence that the cavity performance will be maintained after the string assembly process. This confidence comes from the cavity design that eliminates all indium seals, improved cleaning techniques including high pressure water rinsing, improved cleanrooms, assembly tooling, and procedures as well as demonstrated performance.

Another significant change to the design was moving all the cavity frequency tuner friction bearing components outside the helium and vacuum space. This eliminates motors and gears operating at cryogenic temperatures and in a vacuum. Important features retained from the C20 cryomodule include using all welded

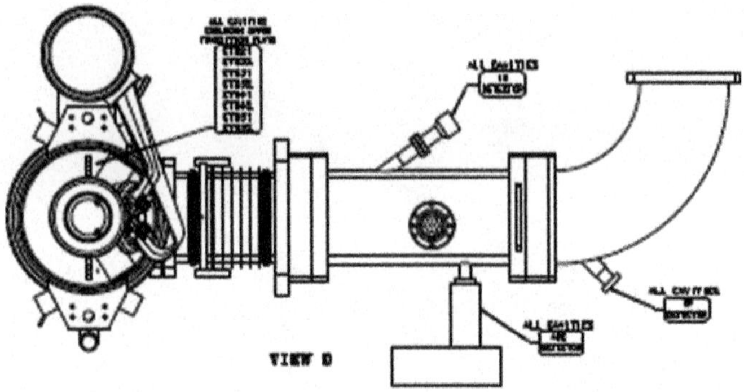

**Fig. 6.24** C100 FPC assembly with the cavity and helium circuits

stainless steel cryogenic circuits including the helium vessel, a double RF window FPC, and individual heaters for each cavity that are inside the helium vessels. The stainless steel cryogenic circuit construction provides a high reliability fabrication avoiding leaks from flanged connections at a minimum cost. Some instrumentation flanges are required to route temperature, level, and heater signals out of the cryogenic circuits. These are kept to a minimum and are done with smaller knife edge seal flanges that undergo rigorous QC checks including cold shocking and leak checking prior to use. The double RF window on the FPC provides redundancy against vacuum leaks that can be a problem on high power RF transmission lines. The space between the two windows creates an intermediate vacuum space that can be monitored for off normal conditions and shutdown RF to avoid catastrophic failures (see Fig. 6.24).

The individual heaters are important for balancing the dissipated power on a cavity by cavity basis on a high power cryomodule and provide for high resolution cavity performance measurements [24] when used in a segmented cryomodule with isolation valves like the CEBAF cryomodules.

### *6.4.3 Cavity*

The SRF cavity is a 1500 MHz seven cell cavity with a close fitting, 10 in. OD, stainless steel helium vessel (see Fig. 6.25). The transition from the Niobium cavity to the stainless steel is done with a brazed assembly incorporated into the beamline just outside of the cavities [25]. The FPC, HOM couplers, and RF pickup probe are located outside of the helium vessel. The cavity assembly includes flanged interfaces for each of these and the two beamline connections. Unlike the original CEBAF cavity where the FPC and HOM and RF pickup probes are located inside the helium vessel, on the C100 cavity these elements are located in the insulating

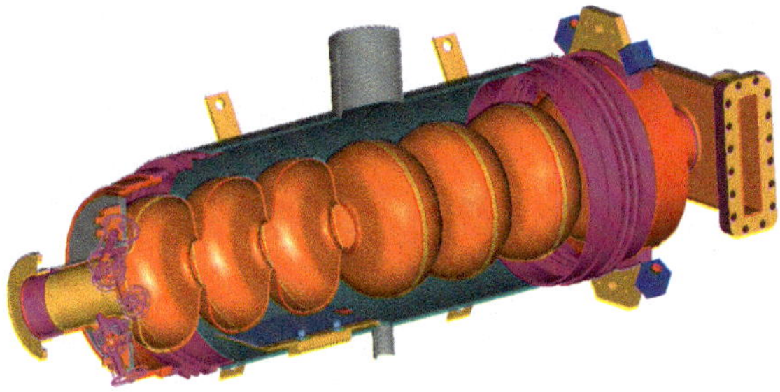

**Fig. 6.25** C100 cavity cut assembly with helium vessel

vacuum space. This requires careful attention to the thermal management [26] as all heat must be conductively removed to the helium inside the helium vessel. The resulting temperature gradient can increase the beamline temperature enough to make them normal conducting. Most often this results in a thermal quench of the cavity. Early on this was a limit in the C100 cavities. The niobium HOM pickup probes would heat from the fundamental mode RF currents and above a certain operating gradient would transition to the normal conducting state and quench the cavity. JLab developed very high thermal conducting RF feedthroughs [27] to shunt the heat from the HOM pickup probe away from the cavity.

### 6.4.4 Cavity Frequency Tuner

The cavity frequency tuner design (see Fig. 6.26) was a radical change from the existing CEBAF tuner [28]. Based on lessons learned from the C20 experience all friction bearing components were placed outside the insulating vacuum space and can be replaced with the cryomodule cold. The tuner actuator has a stepper motor coupled to a harmonic drive reducer for coarse tuning and a piezo actuator for fine tuning. The tuner has a minimum range of 400 kHz and a resolution of 1 Hz. The tuning motion of the drive system enters the cryostat through two thin-wall concentric tubes. The tubes are connected to the upper and lower arms of a scissors type jack. The axial motion of the tubes translates into a linear stroke parallel to the cavity center line. To eliminate friction, motion within the vacuum is done with flexure plates.

### 6.4.5 Cold Mass and Space frame

The C100 cryomodule uses a cold mass subassembly made up of the cavity string (Fig. 6.27), cold magnetic shielding, cavity frequency tuners, cryogenic piping, and associated instrumentation [29]. The cold mass assembly length is constrained to fit into the existing slot length of the original cryomodule while increasing the active

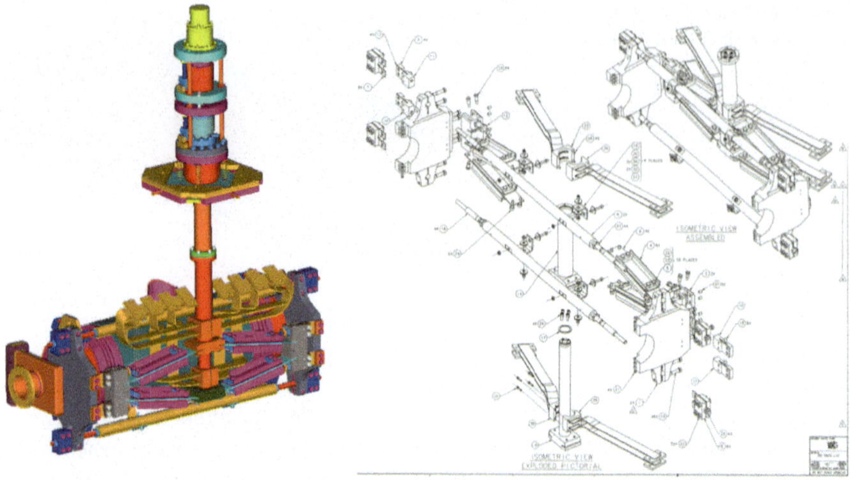

**Fig. 6.26** C100 tuner

**Fig. 6.27** Cavity string anchoring at the center of the space frame using nitronic rods

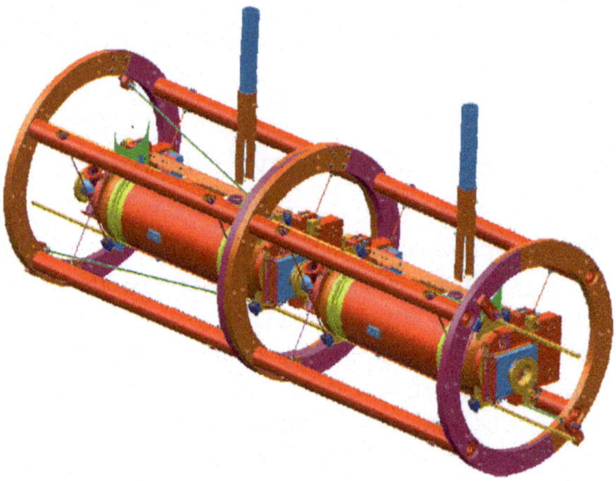

cavity length by 40 %. The increased cavity length is accommodated by the elimination of all bellows between the cavities. The cavity string is anchored in the center of the cryomodule and all cooldown and cavity tuning effects must be managed over the entire length of the four cavity half string extending out from the center to the end of the cryomodule.

There are vacuum boundaries that extend from the cold mass to the warm vacuum vessel at each cavity for the FPC and one at each end of the cavity string (see Fig. 6.28). The relative movement at the cavity string ends is provided for by the use of two multi-convulsion stainless steel bellows in series integrated into the warm to cold beamline transition. There is a shield intercept point between the two bellows minimizing the heat load to the shield and primary cryogenic circuit.

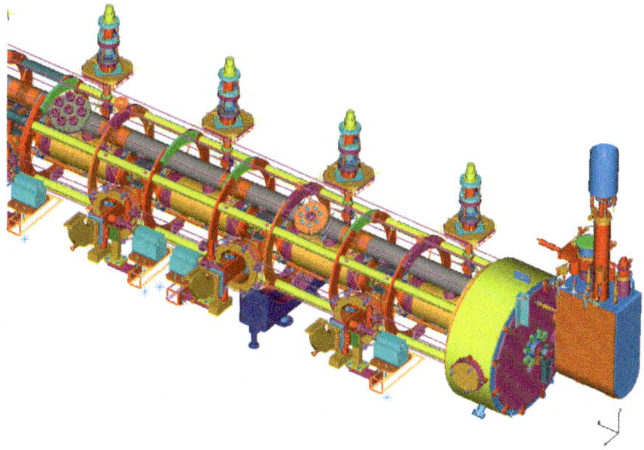

**Fig. 6.28** Cryomodule cut away showing the cold mass and space frame

The FPC is rectangular section mounted normal to the direction of movement which has two 3 convulsion bellows along its length separated as much as possible resulting in a dogleg motion producing the required relative movement between the cavity and vacuum vessel.

The cold mass is supported by the space frame (see Fig. 6.27). The space frame provides advantages over supporting from the vacuum vessel [30]. The first is that it allows all the assembly to take place outside the vacuum vessel (where things are more readily accessible). The space frame with all the cryounits attached, is then rolled into the vacuum vessel and locked down. The second advantage is the requirement for fewer penetrations in the vacuum vessel, thus reducing cost and possibility of vacuum leaks.

The space frame also supplies points for aligning the beam line while inside the vacuum vessel. The space frame consists of a series of five welded tube and ring sections of 304 stainless steel bolted together. The tubes are 2.5 in. outer diameter with a 0.25 in. wall thickness. The rings are 29.75 in. outer diameter, 23.75 inner diameter by 0.75 in. thick. The tubes are fitted to the rings in a socket joint and welded.

The cavities are suspended from the space frame support tubes using nitronic rods. There are eight 0.198 in. diameter rods connected to each unit. The rods have a silicon bronze swaged, threaded fitting attached to each end. The rods are extended to the opposite side of the He vessel in order to increase the length and keep the static heat load to a minimum.

All the mass of the space frame and attachments is supported by 4 stainless steel wheels at the "quarter" points of the frame. Any axial loading (2 g transport loads) is supported by two brackets at one of the quarter points.

### 6.4.6 *Vacuum Vessel*

The vacuum vessel defines the air to insulating vacuum interface. The space frame with the cold mass is inserted axially into the vessel. End caps and tophats are

**Fig. 6.29** Vacuum vessel

installed to complete the cryomodule. The vacuum Vessel consists of 0.25 in. thick 304 stainless steel rolled to a 32 in. outer diameter cylinder. Five sections approximately 61 in. long are welded together using a support ring at each joint and the ends. The support ring is 1 in. thick with a 35 in. outer diameter. There are six 15.5 in. openings and one 17.5 in. opening for waveguide attachment. Also, there are eight 4.5 in. openings for the tuners. Various other small openings for alignment and lockdown access are also included (Fig. 6.29). The vacuum vessel is supported by two saddle stands, one at each "quarter point". The space frame (wheels) is supported by the vacuum vessel at the same points, thus only adding compression stresses into local areas of the wall under normal loading. The space frame is also locked down axially to the vessel via a plate protruding from the space frame, which is trapped in a bracket and welded to the vacuum vessel. There is a spring relief plate located at both ends to limit the internal pressure to less than 2 atm.

## *6.4.7 End Cans*

The End cans for the C100 cryomodule are adapted from the original CEBAF C20 end cans. The primary and shield circuits are connected to the distribution system using removable u-tubes with bayonet connections. Each end can bayonet has an isolation valve allowing u-tubes to be removed and installed while the cryomodule and distribution system are cold. Every volume that can be isolated with valves includes a relief valve to avoid over pressurization in case cold gas is allowed to warm in the potential trapped volume. The supply and return end can piping consists of the primary circuit, which provides cooling fluid to the superconducting cavities, and the shield circuit that cools the nominal 50 K thermal shield. The gas/fluid circuits consist of pipes, tubes, fittings, and valves. The design pressure for the primary circuit is 5.0 and 20.0 atm for the shield circuit [31]. The end plate has a pass through for the beampipe with an O-ring sealing surface as well as a shield thermal strap to minimize the conduction heat load to the 2 K circuit.

### 6.4.8 C100 Performance

Ten C100 cryomodules have been installed and operated in the CEBAF accelerator. Early commissioning of the first cryomodules included high current operations at 108 MV accelerating voltage meeting the design goals. Commissioning results show that these cryomodules will deliver an average energy gain of 110 meV which exceeds the design goal of 108 MV. The C100 cavities are able to operate at an average maximum operating gradient of 19.6 MV/m [32]. Early operations identified challenges to stable operations due to microphonics. As a result the tuner end plates were made stiffer significantly reducing the transmission of vibrations to the cavities.

## 6.5 SSR1 Cryomodule Design for PXIE

### 6.5.1 Introduction

Fermilab is in the process of designing a Project X Injector Experiment (PXIE), a CW linac, to validate the Project X concept, reduce technical risks, and obtain experience in the design and operation of a superconducting proton linac. The overall facility will include an ion source, low and medium-energy beam transport sections, an RF quadrupole, and two superconducting cavity cryomodules. One will contain eight half-wave resonators operating at 162.5 MHz and eight superconducting solenoids. The second will contain eight single spoke resonators (SSR1) operating at 325 MHz and four superconducting solenoids. The design of the cryomodule being developed to house the 325 MHz single spoke resonators and all related systems and services is described here.

The SSR1 cryomodule will operate with continuous wave (CW) RF power and support peak currents of 5 mA chopped with arbitrary patterns to yield an average beam current of 1 mA. The RF coupler design employed should support a future upgrade path with average currents as high as 5 mA. The RF power per cavity at 1 mA average current and 2.2 MV accelerating voltage ($\beta = 0.22$) should not exceed 4 KW with an overhead reserved for microphonics control. The RMS normalized bunch emittance at the CM exit should not exceed 0.25 mm mrad for each of 3 planes.

The current beam optics design for Project X requires that the SSR1 cryomodule contains eight cavities (C) and four solenoids (S) in the following order: C–S–C–C–S–C–C–S–C–C–S–C. Horizontal and vertical dipole corrector are located inside each solenoid. A four-electrode beam position monitor is located at each solenoid.

The intent is that this cryomodule will have all external connections to the cryogenic, RF, and instrumentation systems made at removable junctions at the cryomodule itself. The only connection to the beamline is the beam pipe itself which will be terminated by "particle free" valves at both ends. Minimizing mean

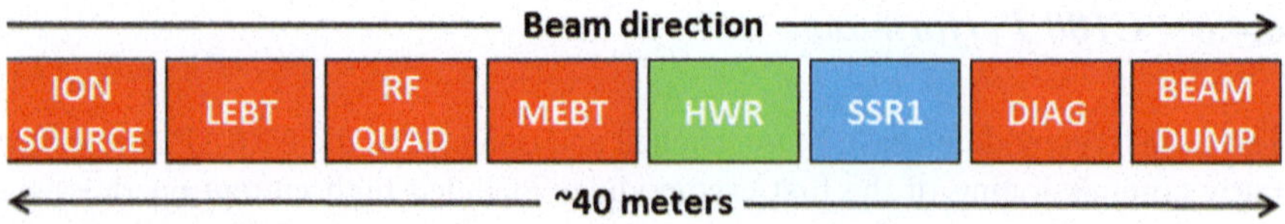


**Fig. 6.30** PXIE layout

time between failure and repair and in situ repair of some internal systems are important design considerations in the cryomodule design. Figure 6.30 shows the linac layout including the location of the SSR1 cryomodule. Table 6.7 lists pertinent operational requirements and parameters of the cryomodule assembly.

## 6.5.2 *Cryomodule Design*

Eventually, Project X will require several different cryomodule designs for cavities operating at 162.5, 325, 650, and 1300 MHz. The SSR1 for PXIE is the first of these being developed at Fermilab. Some details of individual cryomodule components are described in the following sections.

### 6.5.2.1 Cryogenic Systems and Vacuum Interfaces

Fine segmentation is the configuration choice for Project X and PXIE cryomodules. Each individual vacuum vessel will be closed at both ends and the cryogenic circuits will be fed through bayonet connections at each cryomodule. Each cryomodule will have its own connection to the insulating vacuum pumping system. Also, each cryomodule will have its own 2 K heat exchanger and pressure relief line exiting near the middle of the module. This configuration provides flexibility in terms of cryomodule replacement, and cooldown and warm-up times at the expense of requiring more individual cryogenic connections, cold-to-warm transitions at each end of each cryomodule, and extra space at each interconnect to close the beam tube.

### 6.5.2.2 Vacuum Vessel

The vacuum vessel serves to house all the cryomodule components in their as-installed positions, to provide a secure anchor to the tunnel floor, to insulate all cryogenic components in order to minimize heat load to 80, 4.5, and 2 K, as well as maintain the insulating vacuum. It is 1.219 m (48 in.) in diameter and manufactured from carbon steel and is shown as the outermost shell in Fig. 6.37.

**Table 6.7** SSR1 functional requirements specifications (subject to change)

| | |
|---|---|
| General | |
| Physical beam aperture, mm | 30 |
| Overall length (flange-to-flange), m | ≤5.4 |
| Overall width, m | ≤1.6 |
| Beamline height from the floor, m | 1.3 |
| Cryomodule height (from floor), m | ≤2.00 |
| Ceiling height in the tunnel, m | 3.20 |
| Maximum allowed heat load to 70 K, W | 250 |
| Maximum allowed heat load to 5 K, W | 80 |
| Maximum allowed heat load to 2 K, W | 50 |
| Maximum number of lifetime thermal cycles | 50 |
| Intermediate thermal shield temperature, K | 45–80 |
| Thermal intercept temperatures, K | 5 and 45–80 |
| Cryo-system pressure stability at 2 K (RMS), mbar | ~0.1 |
| Environmental contribution to internal field | 15 mG |
| Transverse cavity alignment error, mm RMS | <1 |
| Angular cavity alignment error, mrad RMS | ≤10 |
| Transverse solenoid alignment error, mm RMS | <0.5 |
| Angular solenoid alignment error, mrad RMS | <1 |
| *Cavities* | |
| Number, total | 8 |
| Frequency, MHz | 325 |
| β geometric | 0.22 |
| Operating temperature, K | 2 |
| Operating mode | CW |
| Operating energy gain at β = 0.22, MV/cavity | 2 |
| Coupler type—standard coaxial with impedance, Ω | 105 |
| Coupler power rating, KW | >20 |
| *Solenoids* | |
| Number, total | 4 |
| Operating temperature, K | 2 |
| Current at maximum strength, A | ≤100 |
| $\int B^2 dL$, $T^2m$ | 4.0 |
| Each solenoid has independent powering | |
| *Correctors* | |
| Number, total | 8 |
| Number, per solenoid package | 2 |
| Current, A | ≤50 |
| Strength, T-m | 0.0025 |
| *Beam position monitors* | |
| Number, total | 4 |
| Number of plates | 4 |
| Electrical center accuracy compared to geometric center, mm | ≤±0.5 |

#### 6.5.2.3 Magnetic Shield

Just inside the vacuum vessel, virtually in contact with the inner wall, is a magnetic shield to shield the cavities from the earth's magnetic field. Preliminary tests show that a 1.5 mm-thick mu-metal shield at room temperature reduces the residual field inside the cryostat to less than 10 μT. It is likely that separate magnetic shields will be installed around individual magnetic elements to further reduce the potential for trapped fields in the superconducting cavity structures.

#### 6.5.2.4 Thermal Shield and Multilayer Insulation

Each cryomodule will have a single thermal shield cooled with helium gas, nominally at 45–80 K. It is currently envisioned to be made from 6000-series aluminum with cooling channels on both sides. Two 15-layer blankets of multilayer insulation, between the vacuum vessel and thermal shield will reduce the radiation heat load from the room temperature vacuum vessel to approximately 1.5 W/m2. A 5 K circuit will be available to intercept heat on the input couplers and current leads, but there is no plan to install a full 5 K thermal shield.

#### 6.5.2.5 Support System

All of the cavities and solenoids will be mounted on individual support posts which are in turn mounted to a full-length strongback located between the vacuum vessel and thermal shield. This enables the entire cavity string to be assembled and aligned as a unit then inserted into the vacuum vessel during final assembly. The strongback is aluminum to provide a uniform temperature base. Maintaining the strongback at room temperature helps minimize axial movement of the cold elements during cooldown, reducing displacement of couplers, current leads, and many of the internal piping components.

The support posts are similar to supports utilized in SSC collider dipole magnets (Chap. 2) and ILC and XFEL 1.3 GHz cavity cryomodules (Chap. 5). The main structural element is a glass and epoxy composite tube. The tube ends and any intermediate thermal intercepts are all assembled using conventional shrink-fit assembly techniques in which the composite tube is sandwiched between an outer metal ring and inner metal disk. The strongback and support posts are shown in Fig. 6.31. All of the cavities and magnetic elements are mounted to the support posts using adjustable positioning mechanisms [33].

#### 6.5.2.6 Cavity and Tuner

The cryomodule contains eight single spoke, $\beta = 0.22$, 325 MHz cavities operating in CW mode at 2 K in stainless steel helium vessels. Each has an integral coarse

**Fig. 6.31** Strongback with supports

**Fig. 6.32** Spoke cavity, helium vessel, and tuner

and fine tuner that operates through a lever system and pushes on the cavity end wall. For ease of maintenance, tuner access covers are incorporated into the helium vessel design. The cavity and tuner system is shown in Fig. 6.32 and described more completely in [34].

#### 6.5.2.7 Input Coupler

The input coupler is a 105-ohm coaxial design that supplies approximately 2 kW CW to each cavity in PXIE and ultimately up to 18 kW CW in Project X. The coupler contains a single warm ceramic window that provides separation of the warm and cold coupler sections. During cryomodule fabrication, the cold section can be installed on the cavity in the cleanroom prior to assembly of the string. The warm section can then be installed from outside the vacuum vessel during final assembly. The inner conductor is solid copper with copper bellows to accommodate motion due to misalignment and thermal contraction. The cold end of the outer conductor is 316L-stainless steel. The warm end is copper with copper bellows.

Fig. 6.33 Input coupler

Heat load estimates don't suggest a significant penalty for not copper plating the outer conductor. A forced-air cooling tube is inserted into the inner conductor after assembly that supplies air to cool the coupler tip. Figure 6.33 shows the current coupler design. The input coupler design is described more thoroughly in [35].

#### 6.5.2.8 Current Leads

Each focusing element package contains up to three magnet coils, the main solenoid, operating nominally at 100 A and two steering correctors each operating nominally at 50 A. A conduction cooled current lead design modeled after similar leads installed in the LHC at CERN is being developed for use in the SSR1 cryomodule [36]. Figure 6.34 illustrates the design for the lead assembly. Thermal intercepts at 45–80 K and at 5 K help reduce the heat load to 2 K, nonetheless, these current leads represent a significant source of heat at the low temperature end. There will be one lead assembly for each magnetic element.

#### 6.5.2.9 Solenoid and Beam Position Monitor

The four magnet packages in the cryomodule each contain a focusing solenoid and two dipole correctors all operating in a helium bath at 2 K. The Project X lattice, especially the low-beta section, provides little room along the beamline for beam diagnostics either inside individual cryomodules or between adjacent modules. In order to conserve axial space along the beamline a button-type beam position monitor (BPM) was chosen for installation in the SSR cryomodules. A total of four will be installed in the cryomodule and tested in PXIE, one at each magnetic element. These devices are compact and lend themselves well to incorporation right into the solenoid magnet package as shown below in Fig. 6.35. The bellows in either end of the beam tube allow independent adjustment of each magnet.

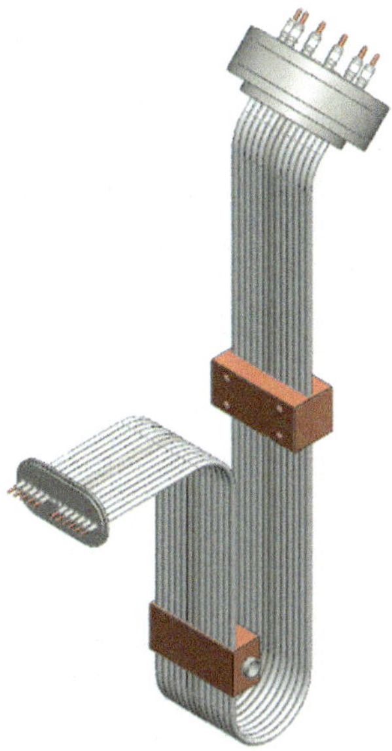

**Fig. 6.34** Conduction cooled current lead assembly

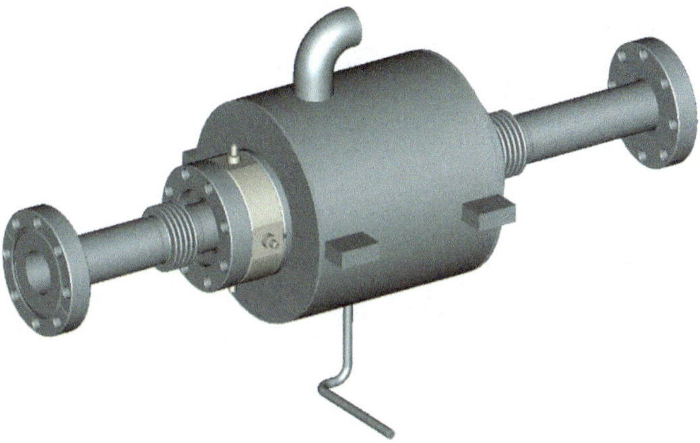

**Fig. 6.35** Solenoid and BPM assembly

### *6.5.3 Final Assembly*

The final assembly of the SSR1 cryomodule for PXIE is shown in Figs. 6.36 and 6.37. Figure 6.36 shows the cavity string consisting of the cavities, solenoids, beam position monitors, and internal piping mounted on support posts which are in turn mounted to the strongback. Figure 6.37 shows the entire cryomodule assembly.

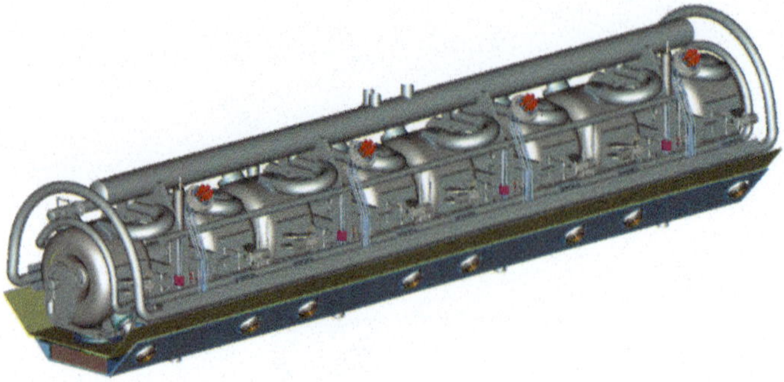

**Fig. 6.36** Cavity string assembly

**Fig. 6.37** Cryomodule assembly

## *6.5.4 Status and Plans*

Fermilab has the PXIE SSR1 vacuum vessel, cavities, strongback, support posts, and many other cryomodule components in-house. Work is in the process to build up production facilities, develop assembly procedures for the cavity string, and to tooling for insertion of the cavity string into the cryomodule. Installation of the cryomodule in the PXIE facility is expected in 2017 [37].

## References


1. G. Oliver, J.P. Thermeau, P. Bosland, G. Devanz, F. Lesigneur, C. Darve, *ESS Cryomodule for Elliptical Cavities*, Proceedings of the 16th International Conference on RF Superconductivity (2013)
2. M. Johnson et al., *Design of the FRIB Cryomodule*, IPAC 2012, New Orleans (2012)
3. H.A. Grunder et al., in *The Continuous Electron Beam Accelerator Facility*, ed. by E.R. Lindsrom, L.S. Taylor. Proceedings of the 1987 IEEE Particle Accelerator Conference, vol. 1 (Washington, D.C., 1987), pp. 13–18
4. C. Reece et al., in *Production Vertical Cavity Pair Testing at CEBAF*, Proceedings of the Particle Accelerator Conference (1993), pp. 650–658
5. C.H. Rode et al., 2.0 K CEBAF CRYOGENICS, CEBAF PR-89-029
6. D. Kashy et al., CEBAF transfer line system, CEBAF PR-91-024 June 1991
7. M. Wiseman et al., CEBAF cryounit loss of vacuum experiment. Appl. Cryog. Technol. **10**, 287–303
8. W. Schneider et al., Thermal performance of the CEBAF superconducting linac cryomodule. Adv. Cryog. Eng. **39**, 589–596
9. J. Benesch et al., CEBAF's SRF cavity manufacturing experience. Adv. Cryog. Eng. **39**, 597–604
10. M. Wiseman, et al, Cryobench—Apparatus for testing cryogenic subcomponents, Appl. Cryog. Technol. **10**, 271–285
11. J.P. Kelley et al., Thermal design and evaluation of the CEBAF superconducting RF cavity's prototype waveguide. Adv. Cryog. Eng. **35**, 675–682 (1990)
12. I.E. Campisi et al., Higher-order-mode damping and microwave absorption at 2 K, EPAC 92, vol. 2
13. G. Ciovati et al., Superconducting prototype cavities for the spallation neutron source (SNS) project. PAC2001, Chicago, IL, June 2001
14. I.E. Campisi et al., The fundamental power coupler for the spallation neutron source (SNS) project. PAC2001, Chicago, IL, June 2001
15. T. Whitlatch et al., Shipping and alignment for the SNS cryomodule. PAC2001, Chicago, IL, June 2001
16. J. Hogan et al., Design of the SNS Cavity Support Structure, PAC2001, Chicago, IL, June 2001
17. T. Whitlatch et al., Shipping and alignment for the SNS Cryomodule. PAC2001, Chicago, IL, June 2001
18. W. Schneider et al., *Thermal Performance of the CEBAF Superconducting Linac Cryomodule*. Advances in Cryogenic Engineering, vol. 39 (Plenum Press, New York, 1994), pp. 589–597
19. D. Kashy et al., CEBAF *Transfer Line Systems*. Advances in Cryogenic Engineering, vol. 37 (Plenum Press, New York, 1992), pp. 577–586
20. L. Cardman, L. Harwood, in *The JLab 12 GeV Energy Upgrade of CEBAF for QCD and Hadronic Physics*, Proceedings of PAC07, pp. 58–62
21. C.E. Reece et al., in *Optimization of the SRF cavity Design for the CEBAF 12 GeV Upgrade*, Proceeding of the 13th International Workshop on RF Superconductivity
22. J. Sekutowicz et al., in *Cavities for JLab's 12 GeV Upgrade*, Proccedings of the 2003 Particle Accelerator Conference
23. D. Douglas, J. Preble, JLAB-TN-98-022
24. M. Drury et al., *Commissioning of the CEBAF Cryomodules*, Proceedings of the 1993 Particle Accelerator Conference, pp. 841–843
25. R. Hicks, JLAB-TN-07-037
26. R. Hicks, E. Daly, JLAB-TN-02-040, *Thermal Analysis of a 13 kW Waveguide for the 12 GeV Upgrade Cryomodule*
27. C. Reece et al., *High Thermal Conductivity Cryogenic RF Feedthroughs for Higher Order Mode Couplers*, Proceedings of 2005 Particle Accelerator Conference

28. K. Davis et al., *Development and Testing of a Prototype Tuner for The CEBAF Upgrade Cryomodule*, Proceedings of the 2001 Particle Accelerator Conference, pp. 1149–1151
29. J. Hogan et al., Design of the CEBAF Energy Upgrade Cryomodule Cold Mass. Proceedings of the 2001 Particle Accelerator Conference, pp. 1595–1597
30. T. Whitlatch, Space frame structural analysis for the CEBAF upgrade cryomodule, JLAB-TN-00-002
31. G. Chang, E. Daly, C100 cryomodule end can piping design pe ASME B31.3, JLAB-TN-07-056
32. M. Drury et al., *CEBAF upgrade: cryomodule performance and lessons learned*, Proceedings of SRF2013, pp. 836–843
33. T.H. Nicol, R.C. Niemann, J.D. Gonczy, *Design and Analysis of the SSC Dipole Magnet Suspension System*, Supercollider 1 (Plenum Press, New York, 1989), pp. 637–649
34. L. Ristori, et al., Design of single spoke resonators for PXIE. Presented at IPAC 2012, paper ID: 2689-WEPPC057
35. S. Kazakov, et al., Main couplers design for project X. Presented at IPAC 2012, paper ID: 2523-WEPPC050
36. A. Ballarino, Conduction-cooled 60 a resistive current leads for LHC dipole correctors, LHC Project Report 691, 2004
37. T.H. Nicol, G. Lanfranco, L. Ristori, High intensity neutrino source superconducting spoke resonator and test cryostat design and status. IEEE Trans. Appl. Supercond. **19**(3), 1432–1435 (2009)

# Chapter 7
# Special Topics in Cryostat Design

**Wolfgang Stautner**

**Abstract** This chapter describes a series of special topics that, while coming from the development of Magnetic Resonance Imaging cryostats, are applicable to many other cryostat designs as well. The topics are: boil off minimization, cryocooler integration, designing with inclined tubes and pressure relief and venting. This chapter contains many figures, tables, equations, design algorithms and references useful to cryostat designers.

## 7.1 Boil off Minimization for Cryostats Without a Cryocooler

This section gives an example calculation for finding the optimal thermal position and length required for linking a thermal shield to a neck tube that supports the helium vessel for minimizing helium boil off (BOFF) [1]. Figure 7.1 shows a typical neck design and Fig. 7.2 shows the model being examined.

The heat balance equation at node T1 is then written as follows:

$$\text{At node T1:}\quad R_{31} - R_{10} - C_{11} + C_{21} - C_{00} + C_{02} = 0 \tag{7.1}$$

The positional variability is indicated by the double arrow at node T1 and X. Note that T3 refers to the temperature of a liquid nitrogen reservoir, usually 80 K.

Where in Fig. 7.2 and Eq. 7.1:

- T1 Node temperature of thermal shield at e.g. 40 K
- T2 Contact temperature of LN2 reservoir or similar with helium vessel suspension tube
- T3 Temperature of liquid nitrogen vessel or thermal shield with temperature T3 (can be higher or lower than T2 to simulate e.g. contact resistance at T2)

W. Stautner (✉)
GE Global Research, Electromagnetics and Superconductivity Lab,
One Research Circle, Niskayuna, NY 12309, USA
e-mail: stautner@research.ge.com

 
J.G. Weisend II (ed.), *Cryostat Design*, International Cryogenics Monograph Series,
DOI 10.1007/978-3-319-31150-0_7

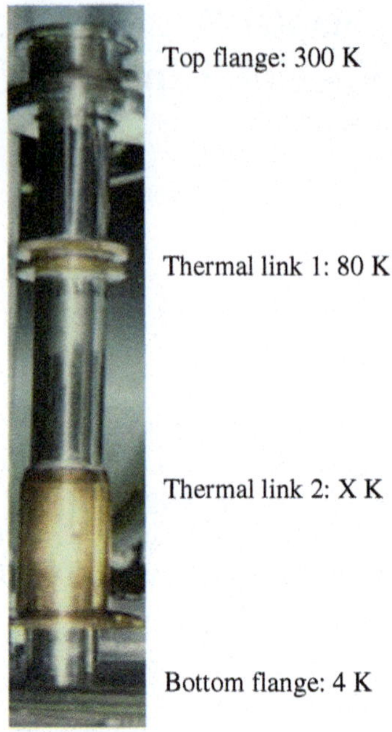


**Fig. 7.1** Typical neck tube with copper links using ultrasonic soldering for consistent thermal contact resistance quality of the copper/steel interfaces

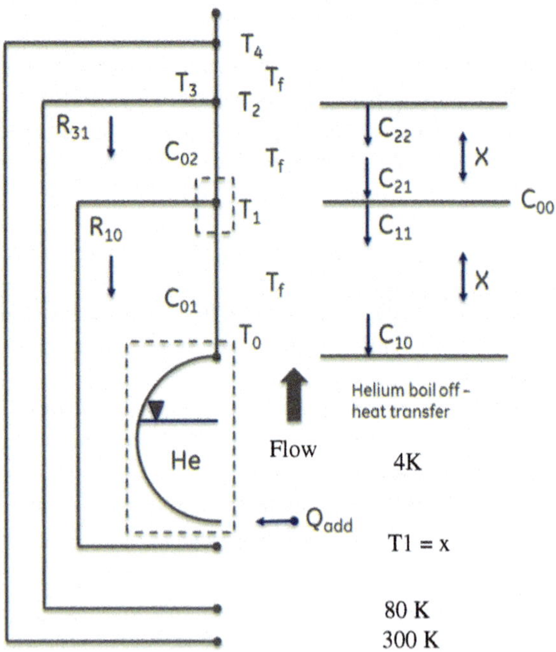


**Fig. 7.2** Modeling heat transport and required contact length to achieve minimum BOFF

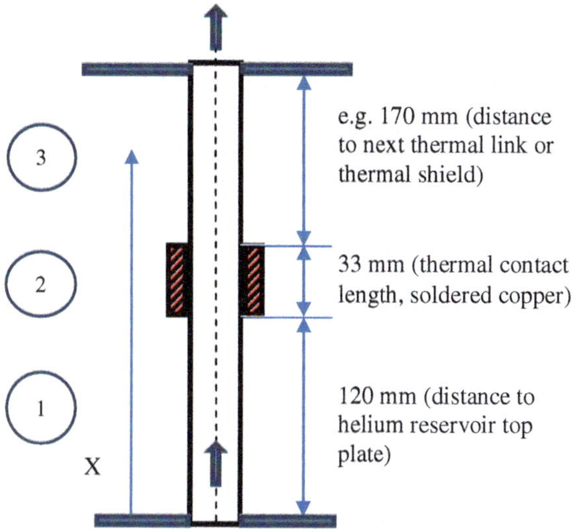


**Fig. 7.3** Tube layout as shown in the example

T4 Temperature of vacuum vessel surrounding liquid nitrogen vessel

$R_{31}$ Thermal radiation from 80 K shield to the intermediate shield temperature

$R_{10}$ Thermal radiation from the intermediate shield temperature to helium reservoir

$C_{22}$ Thermal conduction from liquid nitrogen vessel or thermal shield at this location

$C_{21}$ Thermal conduction at tube bottom into node T1

$C_{11}$ Thermal conduction leaving location T1 entering tube connected to helium vessel

$C_{10}$ Thermal conduction at tube bottom connected to helium vessel

$C_{00}$ Conductive heat loade.g. heat sink to external node leaving thermal boundary at node T1

$C_{01}$ Conduction term due to residual gas (vacuum) to helium vessel (can also be used to simulate contact resistance etc.)

$C_{02}$ Conduction term due to residual gas (vacuum) to intermediate thermal shield (can also be used to simulate contact resistance etc.)

$Q_{add}$ Extra heat term to simulate thermal conduction heat load (e.g. suspensions)

Variables:

$\lambda_{Helium}$ and $\lambda_{Steel}$ = variable thermal conductivities for helium and steel

$cp_{Helium}$ = constant average value between T0 and T1 or T1 and T2

$R_{31} \neq R_{10}$ = thermal radiation

$Nu_1 \neq Nu_2 \neq Nu_3$ = Nusselt number for each tube section

$T3 \neq T2$ = Temperatures can differ

$\varepsilon$ tube 1 $\neq$ $\varepsilon$ tube 2 = tube emissivity coefficient can differ

There are different ways of solving the heat balance equations. A particular useful exercise is the finite difference equation (FDE) approach since this allows you to change the tube structure for each node along the tube length. The equations for conduction, convection, radiation and gas flow as written down by Dusinberre [2, 3] and Croft [4]. The software Kryom Version 3.3 also gives you a good initial estimate on the thermal link position and boil off [5] (Table 7.2).

The energy balance equations solved with FDE for each node are conveniently written as:

$$C_{21} - C_{11} + R_{21} - R_{10} = 0; \tag{7.2}$$

$$C_{10} + C_{add} + R_{10} = H_{lv}; \tag{7.3}$$

$$C_{11} - C_{10} = H_{10} \tag{7.4}$$

$$C_{22} - C_{11} = H_{21}. \tag{7.5}$$

(with conduction and thermal radiation terms as shown in Fig. 7.2 and the increase in enthalpy $H_{10}$ and $H_{21}$ of helium gas and with $H_{lv}$ as the product of latent heat and boiling off helium and considering boil off in ullage space).

We now iteratively search for the best TS link position that results in minimum boil off. Furthermore, we obtain information on thermal shield temperature and the profile along the neck tube (see Fig. 7.4). In addition we also get quantitative results on the individual heat flow parameters for the cryogenic design.

**Table 7.1** Typical neck tube assembly input parameters as shown in Fig. 7.3

| | | | | |
|---|---|---|---|---|
| He − TS − $LN_2$ (K) | 4 | 42 (start value) | 80 | Temperatures |
| Nu1 − Nu2 − Nu3 (–) | 4 | 4 | 4 | Nusselt number |
| dx1 − dx2 − dx3 (mm) | 1 | 1 | 1 | FDE length |
| Positions (1/2/3) | 120 | 33 | 170 | Tube length |
| Ra/Ri (mm) | 15/14.5 | | | Tube radii |
| Emissivity coeff ε (–) | 0.07/0.065 | | | Thermal radiation |
| He − TS − $LN_2$ ($m^2$) | 2.45 | 1.95 | 1.76 | Cryogenic surfaces |

**Table 7.2** Results with Table 7.1 parameters

| | |
|---|---|
| Thermal radiation ($R_{31}$) (mW) | 0.138 |
| Thermal radiation ($R_{10}$) (mW) | 0.0139 |
| Thermal conduction ($C_{10}$) (mW) | 1.02 |
| Thermal conduction ($C_{22}$) (mW) | 127 |
| Thermal conduction ($C_{21}$) (mW) | 18 |
| Thermal conduction ($C_{add}$) (mW) | 1 |
| Optimal TS temperature (K) | 44.6 |
| Optimal contact location (mm) | 120 |
| Optimal TS/Cu contact length (mm) | 33 |
| Minimum boil off per tube (ml/h) | 19 |

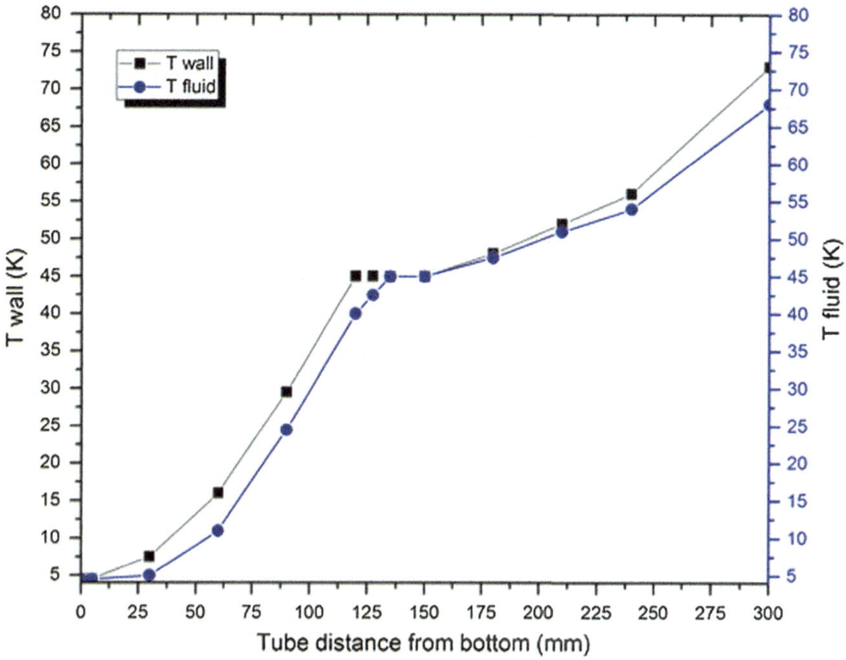


**Fig. 7.4** Optimum thermal link position and required contact length for 290 mm SS tube, 25.4 mm

One can modify the FDE nodes to include annular tubes, corrugated tubes, multiple thermal shields or even forced flow.

## *7.1.1 Discussion*

The process of optimizing boil off of a helium vessel can be explained as follows. Once a thermal link from the TS to the neck tube is made, the heat balance around T1 changes correspondingly until equilibrium between all heat sources is obtained and a continuous boil off rate is established.

Figure 7.5 shows how the parameters within this heat balance node T1 change if the TS link position is moved. If the link, for example, is made higher up the neck tube (Figs. 7.2 and 7.3), the amount of heat conducted down from T2 to T1 increases. As a result, the thermal shield temperature increases as well but since the tube length from T1 to T0 increased boil off will be lower. The designer therefore tries to find the minimum as shown in the curves in Fig. 7.5.

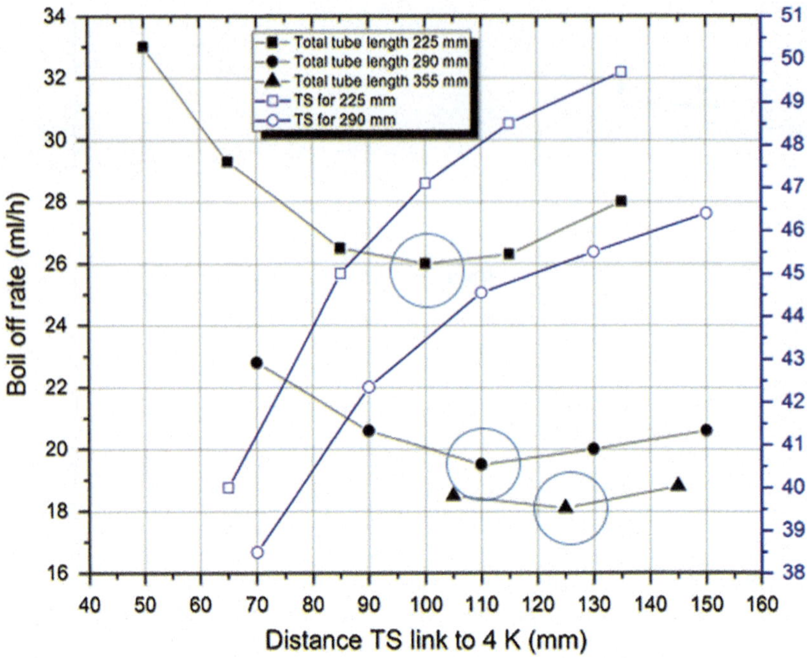


**Fig. 7.5** Optimal thermal shield temperature, link position and minimum BOFF

## 7.1.2 *Pitfalls*

Minimizing boil off is even more important for multiple tubes in contact with a helium vessel. The problem here is to design the tube arrangement in way that all tubes will be receiving the same mass flow rate. In case of flow restrictions at the neck tube bottom (for example a siphon entry tube or a current lead receptacle) the flow along the tubing will not be evenly distributed. The flow restriction can occur in the tube itself in the form of wire looms or copper baffles. Experiments confirmed the existence of large temperature differences between wall and fluid in this case.

For small boil off flow the Nusselt number is almost always 4 [6] which relates to perfect heat transfer between gas and wall. For imperfect heat transfer and non-optimized neck tubes higher boil off rates will be obtained and the temperature difference between wall and gas can become quite large calling for a longer copper link to "lift" up the gas temperature.

The design for minimum boil-off results in efficient cryogen recondensation and re-liquefaction. In times of high cost of helium; re-liquefaction of helium gas, even for smaller cryogenic systems, is convenient and often advisable.

## 7.2 Cryocooler Integration

Since their commercial availability in the 1980s, cryocoolers have become excellent and versatile tools for a diverse range of cryogenic applications, from heat sinking a cold mass, to distributed cooling and even for liquefying or dedicated component cooling. At the cryogenic design stage, the cryogenic engineer needs to know in which way the chosen cooler: Gifford-McMahon (GM), Pulse Tube Refrigerator (PTR) or Stirling type can be integrated in a cryogenic structure and in particular how the thermal links to the cryocooler cold stages should be made.

Generally, the cryocooler interface cost is driven up significantly when demanding minimum temperature difference and by associated requirements, e.g. low vibration transmission, or when it is necessary to retract or remove the cooler. Cryogenic designs have to balance this contact quality effort and make a trade-off against what is needed and is feasible for assembly and cooler maintenance. The most common design solutions for dedicated applications are therefore listed below and illustrated in the following examples.

### *7.2.1 Cryocooler Integration—Options Overview*

Cryocooler attachments for thermal mass cooling

- Fixed — bolt on, braided, internal cryocooler parts removable
- Sleeved — sleeve bolt-on or braided, cryocooler removable from sleeve/serviceable
- Un-sleeved — cryocooler exposed to cryogenic atmosphere, e.g. GM/PTR cooler in helium gas
- Retractable — temporarily engage/disengage from cold mass, e.g. after initial cool-down, sleeved

Cryocooler attachments for specialty applications, distributed cooling and component cooling

- Liquefying coolers (fins machined onto cooler cold stage/recondensing fins in standalone cup)
- Heat-sinked heat pipes
- Heat-sinked current leads
- With heat exchanger (HX) and fan-assisted coolant flow
- Remote cold finger

### 7.2.2 *Cryocooler Integration Examples*

A good example for directly embedding a pulse tube cooler (PTR) in the coil former of the superconducting magnet cold mass itself is the head imaging prototype system built by GE Global Research back in 1995 (see Fig. 7.6). Technically, this is a bold approach since the cooler cannot be removed, but due to the absence of a moving piston no vibrations are transmitted to the superconducting magnet structure. Furthermore, this cooler does not have to be serviced but can be purged of contaminants when warmed up to room temperature, if necessary.

Direct cryocooler cold mass bolt-on is also an interesting option for a number of small-scale MRI magnet systems using medium or high-temperature superconductors (HTS), see also (1) in Fig. 7.14.

Another early, interventional Helium-free LTS MRI magnet system [8] (see Fig. 7.7) built by GE also used fixed installed cryocoolers. Other than PTRs, GM type coolers tend to transmit their piston vibrations to any connected structure. To mitigate this effect flexible braids are introduced between the interfaces to avoid any ghosting appearing in MRI images.

Both coolers in Fig. 7.7 are sleeved to allow the removal of the cooler from the cold plate during service. The cold plate or heat bus extends from the left to the right of both Helmholtz style superconducting $Nb_3Sn$ coils, operating at 10–11 K.

Note that un-sleeved GM coolers can also be serviced if required, by carefully removing the internal piston assembly from the cooler shell tubing. In this case a warm swap to room temperature may be required which requires a cold mass warm up, unless the cooler envelope can be heated to a high enough temperature to allow servicing or refurbishment.

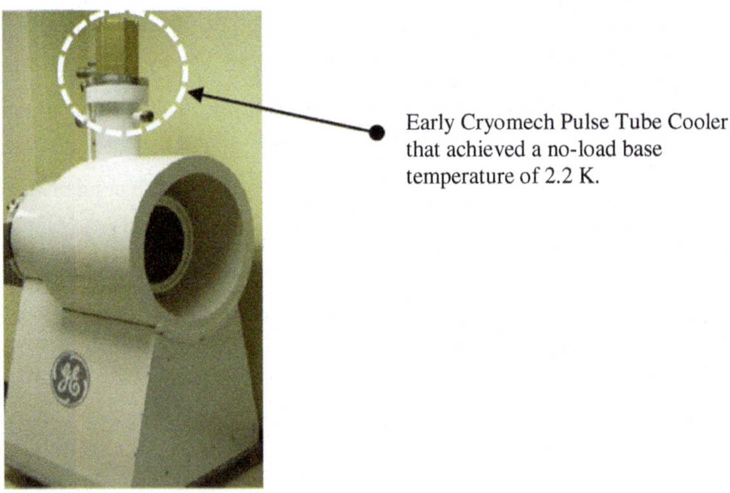


**Fig. 7.6** GE 0.5 T conduction-cooled head imaging system [7]

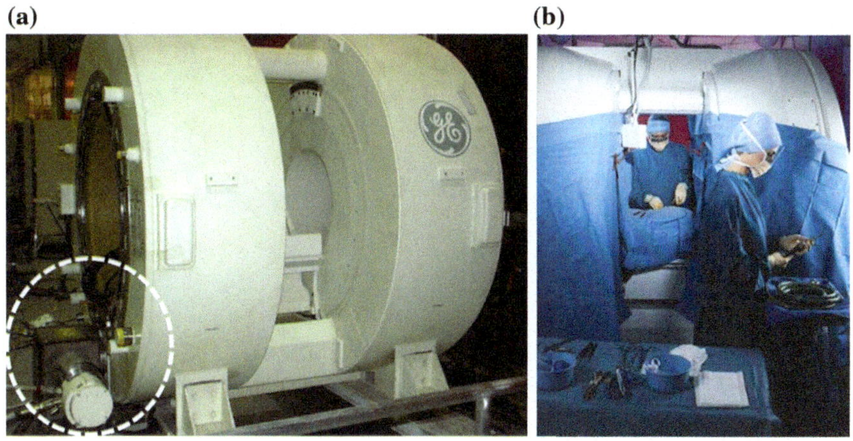


**Fig. 7.7** **a** Cryocoolers (2) (*dotted line*) in cold box and in contact with SC coils via heat bus. **b** Surgeons in the operating theater

### *7.2.3 Schematics and Options of Cryocooler Integration—Overview*

Figure 7.8 shows the most simple and straightforward way of achieving a good thermal contact across the interfaces. The cryocooler comes with bolt holes on the stages for bolting on to a thermal shield or a cold mass. Since the thermal shrinkage at the stages needs to be considered, bolts with Belleville washers are advisable in case of frequent warm up/cooldown.

A typical cryocooler bolt-on configuration for vibration isolation is shown schematically in Fig. 7.9.

For a description of other common flexible conductive links: braids, foils etc., see also [9].

Figure 7.10 shows the race track test coil cooling [10] supported on Heim columns [11, 12] and with attached current leads.

Recently a similar approach has been proposed by Sun [13] for testing race track coils for wind power applications as shown in Fig. 7.11.

All direct bolt-on cryocoolers may be designed sleeved or non-sleeved, depending on maintenance requirements. The same can be said for coolers that are used to recondense cryogens, e.g. helium gas. See Figs. 7.12 and 7.13.

There are downsides to both designs. A non-sleeved design requires a good 1st stage pressed contact onto the cooler interface and a work around for any convective parasitic heat loads between 1st and 2nd stage that may arise (in the absence of stratification). One also has to take precautions against air ingress that later could block internal tubing to an immersed magnet making any recondensation impossible. For the sleeved system, a good, repeatable thermal contact at both cryocooler stages is required.

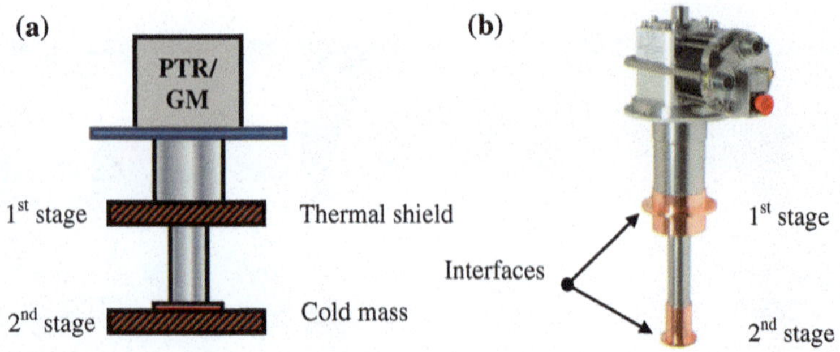


**Fig. 7.8** **a** Cryocooler bolted onto thermal shield and thermal mass, **b** SHI 4 K cryocooler RDK 408 D2 with bolt holes at first and second stages (*courtesy* of SHI cryogenics)

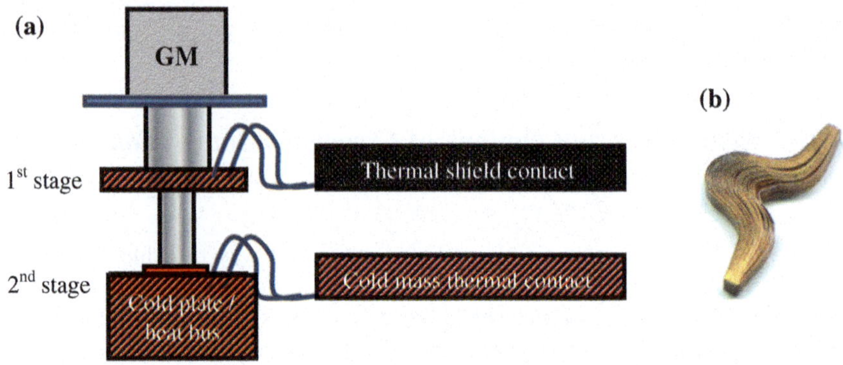


**Fig. 7.9** **a** GM or PTR type cryocooler bolted onto cold plate with highly conductive copper or aluminum braids to cold mass (GM type), see also (*1*) in Fig. 7.14. **b** Example of a foil welded copper link

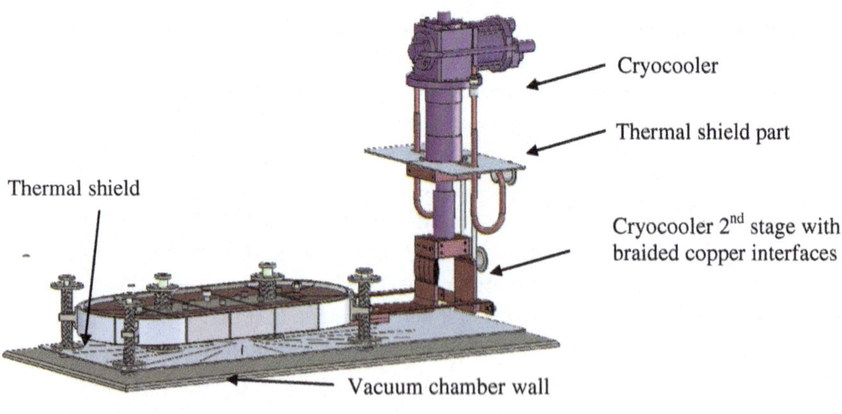


**Fig. 7.10** Bolt-on design (vacuum chamber open for clarity), directly to cold plate and race track cold mass with lead in/out

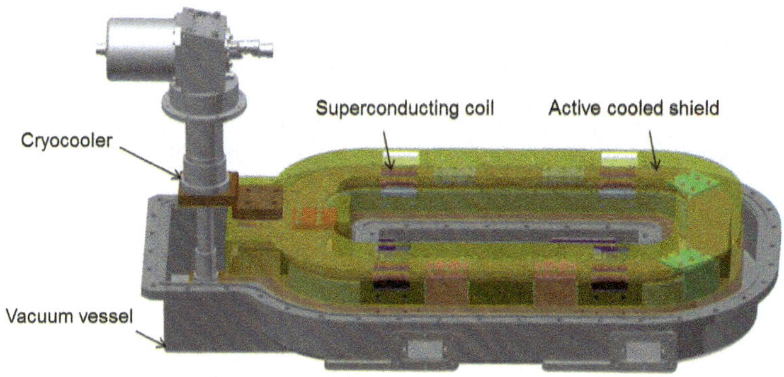


**Fig. 7.11** Race-track coil for wind power applications (test bed)

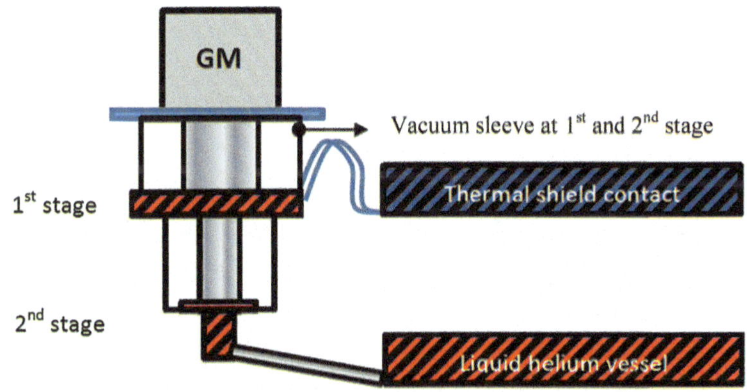


**Fig. 7.12** Sleeved cryocooler and with thermal shield contact with liquefaction cup fitted to the second stage

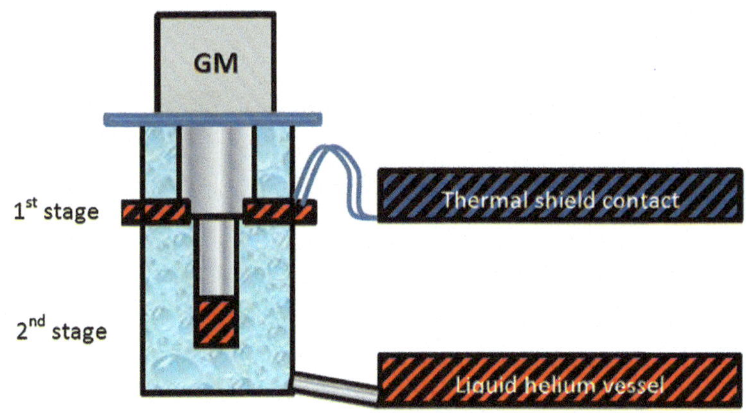


**Fig. 7.13** Sleeveless cryocooler, in direct contact with helium gas and link to thermal shield

Preferred contact means across the interface are either Indium or the less costly Apiezon N® grease.

A cooler can also be integrated and made detachable from its contact stages for a number of reasons. For example, in case of a sudden, unplanned power outage and following compressor shut down resulting in a loss of cooling power, a thermal short would be created that leads to high heat loads to the cooling components (e.g. thermal shield, current leads, cold mass). For large thermal mass cooling (e.g. during cool down) a shielded drive mechanism can be installed. The drive mechanism moves the cooler cold end to the cold mass interface maintaining a specified contact pressure that compensates for the cold mass shrinkage. To avoid a permanent parasitic heat load on the cold mass that can be substantial for larger coolers, the cooler is then retracted automatically [14].

Figure 7.14 shows MRI applications using high temperature superconductors and cooling approaches as discussed in this section.

The cryocooler for the so-called open C-magnet is rigidly and sleeveless bolted onto a cold plate. From the cold plate, braids are routed to the copper sheets that are sandwiched between the individual pancake coils. (2) and (3) show various

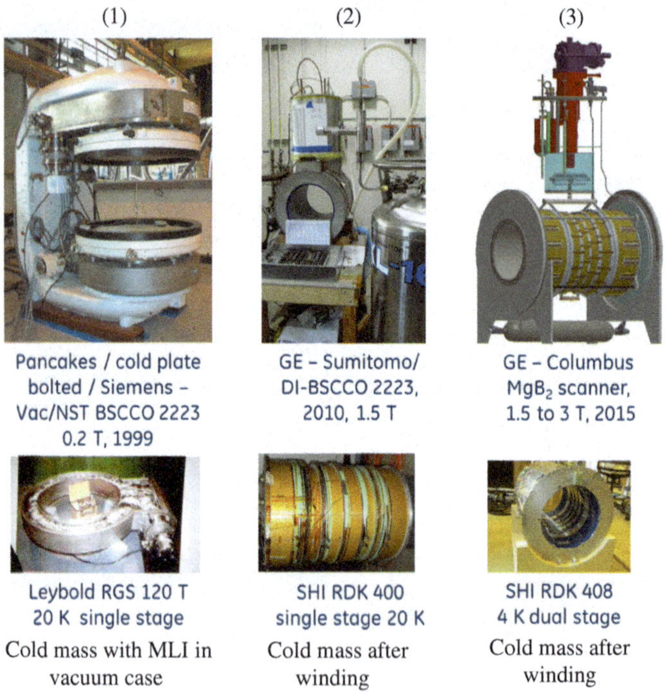


**Fig. 7.14** Cryocooler integration methods for small-scale HTS MRI. *1* Cooling technique: conduction cooling cryogen: n/a cooler: single-stage GM [15]. *2* Cooling technique: thermosiphon cryogen: neon, cooler: single-stage GM [16, 17]. *3* Cooling technique: Thermosiphon, dual cryogens: helium or hydrogen, cooler: dual-stage GM [18]

solutions for extremity scanners for different superconducting wires using thermosiphon technology and cryogen liquefaction with zero boil-off.

### *7.2.4 Cryocooler Integration Techniques for Special Applications*

Heat pipes can be heat-sunk to a first stage of a cryocooler, preferably for sleeved versions. The cold end of the heat pipe extends to the component that needs to be cooled. Heat pipes generally require no maintenance and, depending on the application, are closed system components that give the opportunity of deeply embedding them in a cryogenic structure.

A further common practice is to heat sink the resistive parts of a current lead or the warm end of a HTS lead to a cryocooler. One little complication here is that one needs to work around the conundrum of requiring good thermal contact and electrical isolation that normally exclude each other.

Last but not least several novel applications have surfaced recently [19]. Cryozone for example attaches a heat exchanger onto the cold stage of a cryocooler, mainly for steady state forced-flow HTS application cooling, e.g. superconducting cables etc. [20] (see Fig. 7.15).

It is also possible to run actively cooled circulation loops with hybrid pulse tube refrigerators combined with circulators and cold flow rectifiers as demonstrated by Feller, Salerno, Nellis et al. [21].

Thus, cryocoolers are playing a key role whenever distributed cooling of large surfaces is required. Gas or liquid circulation through cooling loops is possible using micro pumps, blowers or impellers. For HTS applications, the compressed gas can exit, return to the cooler and become part of the distributed cooling circuit.

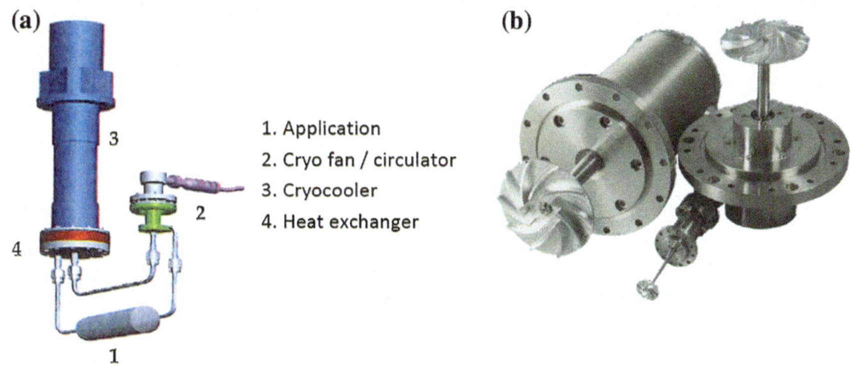


**Fig. 7.15** **a** Cryocooler with attached heat exchanger enabling cryogen flow circulation and **b** cryofans (*courtesy* of Cryozone)

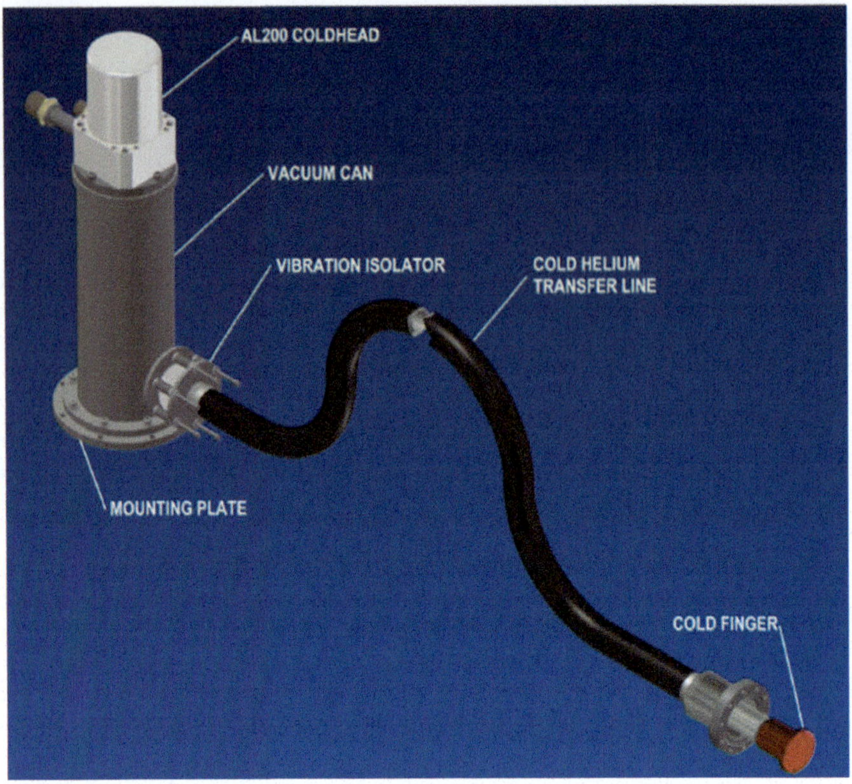


**Fig. 7.16** Cryocooler cold finger cooling with cold helium transfer line (*courtesy* of ICC Press)

Some designs for specific applications require operating the cryocooler at a distance from the object to be cooled. One reason could be the better serviceability for the cooler or if vibration transmission to the object should be minimized which otherwise would interfere with a measurement. This kind of "elongation of the cryocooler cold stage" has recently made possible with a novel approach taken by Cryomech [22] shown in Fig. 7.16. This "cold finger" solution delivers very efficient remote cooling of electronic and other components.

## 7.3 Designing with Inclined Tubes in Cryogenic Systems

When designing cryostats, the engineer occasionally may be tempted to introduce tubing, penetrations or gas filled sleeves in an angled orientation. There are a number of reasons for implementing those, either for getting better accessibility to the internal cryostat structure from the top, when patient room height limitations need to be considered or simply for better liquid draining.

In other designs where the cryostat envelope space is tight, gas filled tubes and/or operating under high pressure can only be fitted inclined. Furthermore, cryocoolers themselves are gas filled tubes under pressure, in particular the pulse tube cooler, with its absence of any solid moving piston. The moving piston here is the oscillating gas column itself. Not surprisingly there are operational limits due to inclination for this cooler, as we will see.

So why should the cryogenic designer be careful when routing inclined tubes or a network of pipes? Inclined tubes generally introduce a parasitic heat flow driven by the density difference in the working medium due to the difference in the length of the wall conduction heat path and the resulting heat load that sets up fluid motion within these tubes. If there is only a small deviation required from the vertical it is almost always better to keep the tubing vertical. The answer really depends on how critical it is to maintain the cryostat heat balance and whether any cost increase that may result is acceptable.

Inclined tubes can be classified as follows:

- Inclined tubes (neck tubes/turrets) at low or high operating pressure (open[1]/closed)
- Inclined tubes with pulsating flow at high/low frequency
- Inclined tubes for thermal mass cool-down
- Inclined heat pipes
- Inclined thermosiphons

When designing cryostats we often notice that the heat balance deviates from expected, calculated values. It is then we look out for parasitic heat loads that may have been missed in the design or at last minute changes in the assembly. As Scurlock noted in 1977 [23], "the boil-off is always 50–100 % greater than the design figures". This prompted Islam and Scurlock to conduct vapor column flow visualization experiments for vertical and inclined tubes. The team was the first to show in which way the gas column flow changes when tubes are tilted. Tests were performed on wide open tubes as well as on ones with small diameters that are open to atmosphere. It was possible to differentiate between 6 distinct regions in which the flow changes within the column.

Figure 7.17 shows a typical tubing that gives easy access to a liquid reservoir, e.g. for current lead insertion or for filling the reservoir or for using it as a boil-off tube.

This tilt angle was identified as a source of heat leak into the cryogenic liquid storage vessel. Further research detail is given in [24–28]. With the starting interest in pulse tube cooler technology researchers quickly noticed the orientation dependency of the tubing. A typical curve is shown by Kasturirengan [29], (see Fig. 7.18a) for a pulse tube cooler running at 15 bar at a frequency of 1.6 Hz. The inclination angle is shown in Fig. 7.18b. Figure 7.18a shows the convective, or

[1]Open under high pressure in this case means the open end of the tube is connected to a larger gas reservoir under high pressure.

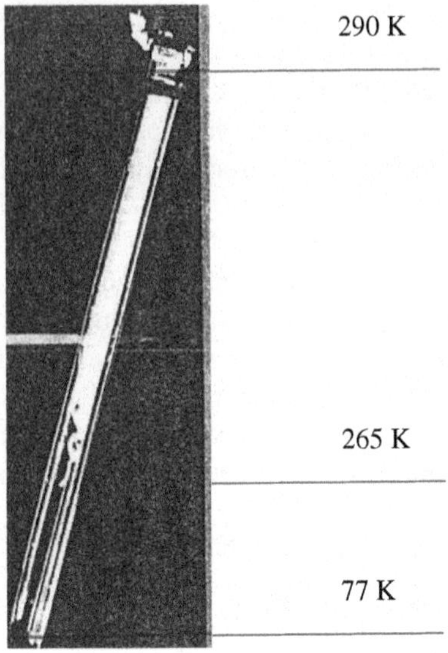


Fig. 7.17 Tilted, open access tube for a cryostat [23]

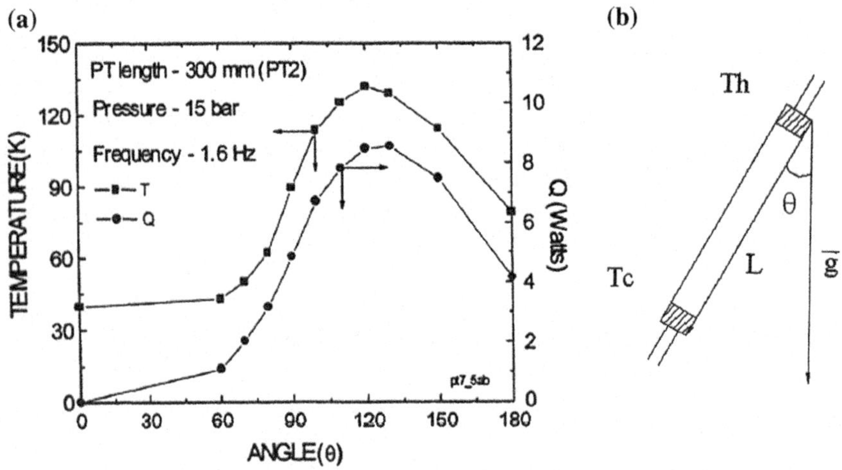


**Fig. 7.18** **a** Typical PTR cooling power loss, **b** at different inclination angles [29]

parasitic heat loss that reduces the cooling power at different inclination angles. For high frequency operation this effect is reduced as Thummes reports [30].

The recent research of Langebach and Haberstroh [31] and others is of great value when inserting tubes in cryostats that work under high pressure and are closed or open (closed with reservoirs) at the ends. Why is that important for cryostat designs? Cryostats sometimes are not only used as storage container or for experiments but for housing a cold mass and in particular a superconducting

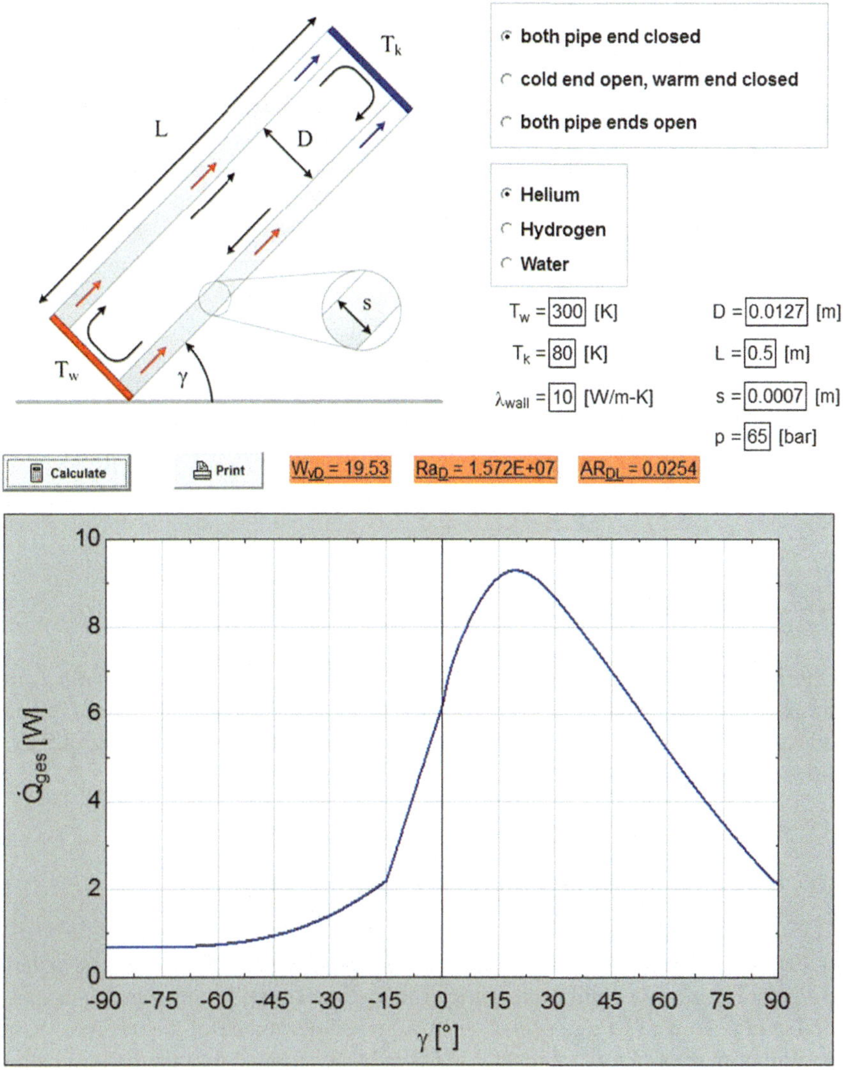


**Fig. 7.19** Tilted pipe under pressure—convective behavior (*courtesy* of Langebach)

magnet. The cool-down of a cold mass is often facilitated and accelerated using a heat pipe, a typical, early configuration is given by Begulilowa [32]. In cryogenic designs, very rarely has one the comfort of routing the heat pipe vertically and they most likely will always be inclined. For those instances the computer program developed by Langebach [33] (based on his experimental correlations) calculates the performance loss due to the inclination effect before the heat pipe starts to liquefy at the cold end that is, e.g. during cool-down from room temperature to its operating temperature. Once liquid is formed within the tube and the pressure within the tube is reduced to its designed operating pressure other curves apply that show the degradation in performance of an unwicked heat pipe at tilted angles.

Referring to Fig. 7.19 −90° defines the tube with its warm end up ($T_w$) whereas +90° means the cold end is up ($T_k$). Input parameters are operating pressure in bar, temperature and average thermal conductivity of the tube wall, length, diameter and thickness. Further information is required on whether the tube is closed at both ends or only at one end and what cryogen is used (helium, hydrogen, nitrogen). In his thesis Langebach examines different tubes at different boundary conditions and Raleigh numbers and gives a range of methods to prevent convection in inclined tubes from happening. With one of his solutions—the "snake" tube—parasitic heat loads were considerably reduced.

As for the particular use of an inclined thermosiphon tubing, this highlights an interesting aspect that needs further investigation. For some applications, boundary conditions and operating conditions it may be beneficial to incline a thermosiphon tube whenever a high quality thermal short is required, e.g. during an initial cool-down.

### *7.3.1 Pitfalls*

A horizontal thermosiphon tubing or a horizontal manifold is known not to work and should be avoided as it stops the bubble movement (gas bubble entrapment) and causes flow instabilities. For bubbles to move towards a recondensing surface, a horizontal inclination angle of at least 5° is required.

## 7.4 Cryogenics for Cryostats: Pressure Rise

All cryostats, such as those used for accelerator magnets, SC generators or large scale storage vessels require design against overpressure. In the event of a superconducting magnet "going normal" or by a preceding rupture of the vacuum vessel, the stored magnet energy of several MJ is quickly dissipated to the helium bath as well as to the magnet former itself (Fig. 7.20).

The amount of heat transferred to the bath is a function of the magnet energy content, the current decay time constant, the transient heat transfer to liquid helium (the obtainable heat flux density (see Table 7.3) in W/cm$^2$), vessel volume and fill level, the set pressure for the relief valve or burst disk, the magnet former surface structure, as well as type and orientation of the latter. Each liter of liquid helium at 4 K creates 700 L of gaseous helium that need to be discharged safely; for a 3 T magnet cryomodule this would amount to 1.4 million gas liters. External quench gas capture is therefore difficult. Using a cryocooler for reclaiming the quench gas and associated cooldown usually is not feasible within a reasonable timeframe for standard bath-cooled cryomodules. If possible, transfer liquid helium back into

**Fig. 7.20** Quench gas release through quench duct after decommissioning a cryomodule for MRI (MRI magnet installed in hospital basement)

**Table 7.3** Typical heat flux densities in W/cm$^2$ for cryostats (collapse and/or quench) [34]

| Designs—typical configuration | Flux |
|---|---|
| Superinsulated bath cryostat | 0.6 |
| Liquid helium transport can with gas-cooled shields | 2.0 |
| Helium vessel without insulation | 3.8 |
| Usually for quench calculations | 4–5 |

dewars rather releasing helium to outer space when decommissioning complete cryomodule systems.

### 7.4.1 *Quench Pressure Rise in Cryostats and Quench Duct Sizing—A Modeling Example*

Predicting quench pressure rises in cryomodules is generally a difficult task, but nevertheless necessary. The possible increase in 4 K magnet temperature to 40–50 K after a quench—depends on the chosen coil structure, the coil former material and mass as well as on the amount and type of wire and coil configuration (e.g. race track coil or solenoid). When designing the cryostat the important operating conditions have to be fulfilled, with the magnet quench posing all greatest challenge. Peak mass flow rates of up to 5 kg/s leaving the cryostat at a temperature of 10 K, partly accompanied by liquid helium expelled at the turret exit (see position 2 in Fig. 7.21) are not unheard of.

Quench pressure rise modeling has been attempted previously with forced flow fusion type magnets [35, 36], accelerator quadrupole magnets [37] and smaller solenoids [38], but only very little so far has been done on many magnet systems, see for example in [39]. An excellent attempt in how to solve the similar cryostat

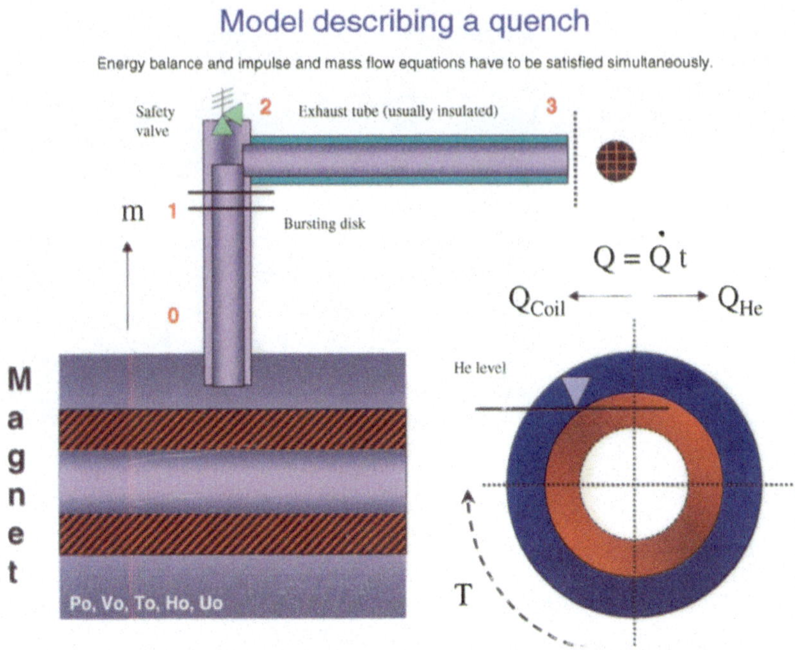


**Fig. 7.21** Basic quench parameters for state-of-the-art cryostats and energy transfer

vacuum collapse problem has been pointed out by the late Walker [40] on a 19,000 L helium storage tank at Fermi lab.

For a magnet system, pressure P, volume V, temperature T, enthalpy H and internal energy I at the initial state 0 are given. Once the magnet dissipates its energy to coil former and helium bath, the pressure rise can be described as isochoric, followed by an isentropic expansion [41] through the valve system (Fig. 7.21: position 2) once the valve opens, and for reasons of simplicity the flow can be regarded as adiabatic through the quench duct. The stepwise iterative calculation with real gas properties is rather complex involving compressible friction flow close to the speed of sound simultaneously satisfying gas dynamic equations for Helium (Fanno line) with the associated transient heat transfer. Figure 7.22b shows a theoretical quench pressure calculation result. Green line (bar) = quench pressure decay (dip after burst disk opens), black (kg/s) = volume flow out of cryostat, red (K) = temperature rise of the helium vessel over 6 s. Safe discharging of quench gas requires careful design of the turret diameter and providing and maintaining an unblocked gas passage. The gas transfer mechanism from the cryostat to ambient atmosphere requires a reliable safety valve, a burst disk and a quench gas line of sufficient diameter, the latter depending on the customer siting requirements. Figure 7.22a shows typical quench rates in state-of-the-art high-energy magnets in cryomodules. Note the sudden rise in pressure from 0 to 5 psi within 2 s. The entire duration of the gas release depends mainly on the

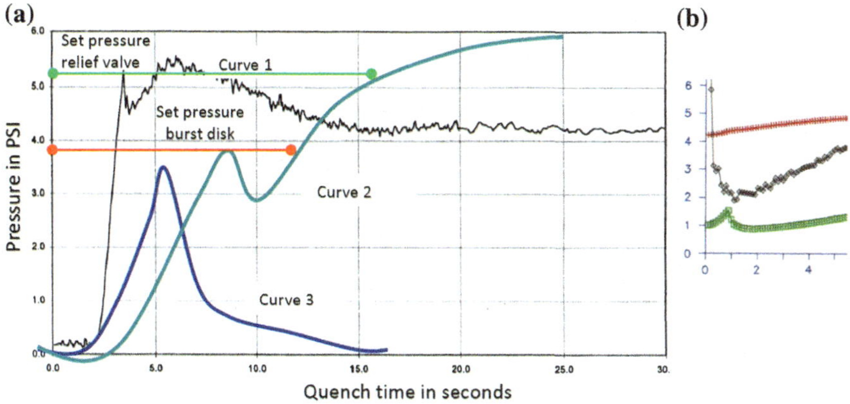


**Fig. 7.22** **a** Typical examples of quench test results for cryo-modules cryostats, **b** modeling

opening diameter of the quench turret and on the discharging rate of the cryostat valve. In the example in Fig. 7.22a, curve 1, the total magnet quench time is approx. 5 s; after that time, the pressure and mass flow starts to decay exponentially.

Curve 1 represents the course of the quench pressure when selecting a relief valve that could cause potential problems for several reasons: although the quench valve opens and the pressure drops immediately, the quench pressure still rises again 5 s after the relief valve opens and remains constant. Three problems are highlighted with this example. Firstly, should the gas flow see a further restriction, e.g. due to ice formation, the pressure increase could make the burst disk blow which is undesirable since then the magnet is exposed to ingress of air and further turret ice-up. Secondly, in case the flow restriction in the safety valve increases even further and, if the burst disk is chosen with a too small a diameter the quench pressure itself can exceed the design pressure of the cryostat and invalidate the coding requirements for this cryostat vessel. Lastly, the long quench duration itself leads to icing up of further vent line components and faults. Design guidelines are given in [42, 43]. Curves 2 and 3 show the course of the quench pressure curve for a burst disk opening pressure of 4 psi (red line). Curve 3 safely releases the helium gas with sufficient discharge rate and a high enough kV-value, whereas curve 2 compromises the pressure vessel design by considerably exceeding the design pressure with the danger of rupturing the helium vessel or fitted components. It is also interesting to note that the opening diameter initially helps to quickly release gas into the quench duct, hence the dip in pressure. There is however a sharp increase one needs to watch out for when fitting long quench ducts. Curve 2 is also a typical example where the quench duct is too long leading to a back pressure in the cryostat. In this case the diameter has to be increased or the number of pipe bends reduced.

## 7.5 Advanced Cryostat Cryogenics—Carbon Footprint Considerations

"Advanced cryogenics" [44] aims to reduce the liquid helium inventory by introducing heat pipe technology, creating a basically "near-dry" type, conduction-cooled magnet as introduced by the accelerator magnet industry in the 1970s. One of the downsides of a reduced cryogenic volume is the longer time needed to recool the magnet back to operating temperature. On a more positive note any loss of vacuum in the cryostat triggering a quench for example, no longer requires satisfying the high safety ordinance efforts as required for a 3000 L helium vessel.

## References


1. W. Stautner, Quantitative energy balance analysis in cryostats, in *Physicist's Conference, Oral presentation*, Münster, 1984
2. G. M. Dusinberre, *Heat Transfer Calculation by Finite Differences* (International Text book Company, 1960)
3. G.M. Dusinberre, *Numerical analysis of heat flow* (McGraw Hill, 1949)
4. D. Croft, D. Lilley, *Heat Transfer Calculations Using Finite Difference Equations* (Applied Science Publishers, 1977)
5. P. Hanzelka, I. Vejchoda, Academy of sciences of the Czech Republic, Institute of Scientific Instruments (2003)
6. G. Arahonian, L.G. Hyman, L. Roberts, Behavior of power leads for superconducting magnets. Cryogenics **21**, 145 (1981)
7. J.M. van Oort, E.T. Laskaris, P.S. Thompson, B. Dorri, K.G. Herd, A cryogen-free 0.5 Tesla MRI magnet for head imaging. Adv. Cryog. Eng. **43**, 139–147 (1998)
8. K.G. Herd, E.T. Laskaris, P.S. Thompson, A dual refrigerator assembly for cryogen-free superconducting magnet applications. IEEE Trans. Appl. Supercond. **5**, 185–188 (1995)
9. J R. Ross, M. Donabedian, Spacecraft thermal control handbook. Cryogenics **II** (2003)
10. W. Stautner, K. Sivasubramaniam, E.T. Laskaris, S. Mine, J. Rochford, E. Budesheim, K. Amm, A cryo-free 10 T high-field magnet system for a novel superconducting application. IEEE Trans. Appl. Supercond. **21**, 2225–2228 (2011)
11. J.R. Heim, The heim column, National accelerator laboratory, report TM.334A (1971), pp. 1–21
12. G. Hartwig, Support elements with extremely negative thermal expansion. Cryogenics **35**, 717–718 (1995)
13. J. Sun, S. Sanz, H. Neumann, Conceptual design and thermal analysis of a modular cryostat for one single coil of a 10 MW offshore superconducting wind turbine, in *IOP Conference Series*, vol. 101 (2015) p. 012088
14. W. Stautner, Remote actuated cryocooler for SC generator and method of assembly the same, US20140100113A1
15. F.J. Davies, W. Stautner, A.F. Byrne, M. Wilson *An HTS magnet for whole-body MRI*, EUCAS'99
16. W. Stautner, K. Amm, E.T. Laskaris, M. Xu, X. Huang, A new cooling technology for the cooling of HTS magnets. IEEE Trans. Appl. Supercond. **17**, 2200–2203 (2007)

17. W. Stautner, M. Xu, E.T. Laskaris, G. Conte, P.S. Thompson, C. van Epps, K. Amm, The cryogenics of a thermosiphon-cooled HTS MRI magnet—assembly and component testing. IEEE Trans. Appl. Supercond. **21**, 2096–2098 (2011)
18. W. Stautner, M. Xu, S. Mine, K. Amm, Hydrogen cooling options for MgB2-based superconducting systems, in *AIP Conference Proceedings*, vol. 1573 (2014), p. 82
19. W. Stautner. K. Amm, M. Xu, Cooling systems for HTS applications—overview and critical assessment, IWC-HTS plenary talk 1, Matsue-Shi (2015)
20. H. Vermeulen, Cryogenic circulators: the solution for cooling problems? Cold Facts **29**(2), 49–48 (2013)
21. K.R. Feller, L.J. Salerno, A. Kashani, B.P. Helvensteijn, J.R. Maddocks, G.F. Nellis, Y. B. Gianchandani, Technologies for cooling of large distributed loads, AIAAA, 092497 (2008)
22. C. Wang, E. Brown, A. Friebel, A compact cold helium circulation system with GM cryocooler, in *18th International cryocooler conference ICC*, Syracuse (2014)
23. M.S. Islam, R.G. Scurlock, Qualitative details of the complex flow in cryogenic vapor columns. Cryogenics 655–680 (1977)
24. P. Lnyam, A.M. Mustafa, W. Proctor, R.G. Scurlock, Reduction of the heat flux into liquid helium in wide necked metal dewars. Cryogenics 242–247 (1969)
25. M.S. Islam, R.G. Scurlock, Analysis of solid vapor heat transfer in helium vapor columns at low temperatures. Cryogenics 323–328 (1978)
26. M.S. Islam, D.J. Richards, R.G. Scurlock, The influences of thermal stratification and flow interaction on the enhanced natural convective heat transfer at low temperatures. Cryogenics 319–325 (1978)
27. J. Boarman, P. Lynam, R.G. Scurlock, Complex flow in vapor columns over boiling liquids. Cryogenics 520–523 (1973)
28. P. Lnyam, W. Proctor, R.G. Scurlock, Reduction of the evaporation rate of liquid helium in wide necked dewars, in *Heat Flow Below 100 K*, no. 2 (International Institute of Refrigeration, Paris, France, 1965), pp 351–247
29. S. Kasturirengan, S. Jacob et al., Experimental studies of convection in a single stage pulse tube refrigerator. Adv. Cryog. Eng. **49**, 1474–1481 (2003)
30. G. Thummes et al., Convective heat losses in pulse tube coolers: effect of pulse tube inclination. Cryocoolers **9**, 393–402 (1997)
31. R. Langebach, C. Haberstroh, Natural convection in inclined pipes—a new correlation for heat transfer estimations, in *AIP Conference Proceedings*, vol. **1573** (2014), pp. 1504–1511
32. R. Bewilogua et al., Application of the thermosiphon for precooling apparatus. Cryogenics **6**, 34–36 (1966)
33. R. Langebach, Wärmeeintrag durch geneigte Rohrleitungen in kryogene Speicherbehälter, Dissertation, TUD press, 2013, URL: http://books.google.de/books?id=Cez5nAEACAAJ
34. W. Lehmann, Internal report, Safety aspects LHe cryostats and LHe transport containers, Research Center Karlsruhe, Report 08.01.01P04B (1978)
35. J.R. Miller, ORNL, Pressure rise during the quench of a superconducting magnet using internally cooled conductors (1980), pp. 321–329
36. P.H. Eberhard, et al, Lawrence Berkeley Lab, Quenches in large superconducting magnets, in *Proceedings of 6th international Conference on Magazine Technology (MT 6)*, Paper 75, Bratislava (1977), pp. 654–662
37. K. N. Henrichsen, et al, Analysis of some resistive transitions in the ISR super-conducting quadrupole magnets. Adv. Cryog. Eng. **27**, 245–256 (1982)
38. V. Kadambi, B. Dorri, Current decay and temperatures during superconducting magnet coil quench. Cryogenics 157–164 (1986)
39. B. Seeber, Handbook of applied superconductivity, in *Pressure increase during a quench*, Vol. 2, Figure G2.2.24 (Stautner) (1998), p. 1235
40. R.J. Walker, Calculation of the pressure rise in the Fermilab 19000 l helium dewar, Adv. Cryog. Eng. **29**, 777–784
41. R.J. Walker, private communication (1985)

42. G. Bozóki, *Überdrucksicherungen für Behälter und Rohrleitungen*. Verlag TÜV Rheinland (1977)
43. W. Lehmann, *Sicherheitsauflagen beim Engineering von LHe- und LN2-Apparaten und – Anlagen,* Research Center Karlsruhe, Report 03.05.01P02A (1982)
44. Y. Lvovsky, W. Stautner, Novel technologies and configurations of superconducting magnets for MRI. Supercond. Sci. Technol. **26**(9), article id. 093001 (2013)

# Chapter 8
# Design and Operation of a Large, Low Background, 50 mK Cryostat for the Cryogenic Dark Matter Search

**Richard L. Schmitt**

**Abstract** Cryostats that operate below 1 K have additional requirements involving the need for extremely small heat leaks and alternative cooling methods. This chapter describes the design and operation of a cryostat operating at 50 mK for the Cryogenic Dark Matter Search experiment. Included are descriptions of the cooling and thermal insulation system, seal design, fabrication and operations. Valuable data is provided on the thermal conductivity of Kevlar and the calculation of joint conductance. The particular issue of using only radiopure materials in the cryostat construction is also covered. A list of lessons learned from the cryostat operation is provided.

## 8.1 Introduction

To support the Cryogenic Dark Matter Search (CDMS) experiment, a sub Kelvin cryostat and support system was built and operated at Soudan MN. In addition to the challenges of milliKelvin operation, the cryostat was made with low background material, and was located in an RF shielded clean room one-half mile below the surface at Soudan Underground State Park.

## 8.2 Physics Detectors and Towers

The goal of CDMS was to detect dark matter, which has been detected through large-scale gravitational interactions. CDMS used cryogenic germanium and silicon detectors, which are capable of detecting weakly interactive dark matter (WIMPs).

WIMPs are detected through their interactions with the nuclei in the germanium. When a nucleus is hit, it recoils, causing the whole germanium crystal to vibrate.

---

R.L. Schmitt (✉)
Fermi National Accelerator Lab, P.O. Box 500, Batavia, IL 60510, USA
e-mail: rlschmitt@fnal.gov

 
J.G. Weisend II (ed.), *Cryostat Design*, International Cryogenics Monograph Series,
DOI 10.1007/978-3-319-31150-0_8

These vibrations, or phonons, propagate to the surface of the crystal where they heat sensors consisting of thin aluminum traps connected by tungsten meanders. The tungsten is kept at its Tc (critical temperature) and so is an ultra-sensitive thermometer, one capable of sensing temperature changes a fraction of a milliKelvin. The extra energy—as little as 5 keV—the phonons bring raises the tungsten's resistance by a minute amount. When the phonons reach the aluminum, they excite quasi-particle states, which propagate to the tungsten and heat it up.

When the temperature of the tungsten rises so does the resistance of the circuit. This causes the bias current to decrease since the voltage across the tungsten is held constant. The resulting pulse is picked up by SQUID (Superconducting QUantum Interference Device) amplifiers.

There are other particles that go through the detectors besides WIMPs. The CDMS detectors were shielded to minimize the number of these other particles. The detectors are capable of discriminating between most of them and the WIMP signal [3, 17, 22].

The detectors ran in the Soudan Mine in Minnesota, a half-mile underground. The deep site is chosen to shield the detectors from cosmic rays.

Six detector crystals were mounted in each tower assembly (see Fig. 8.1). The towers provided structural support for the detectors and thermal heat sinking at the base temperature. The tower features included heat sinks at 200 mK, 1 and 5 K for the structural supports and signal wiring. Wiring and structural elements in the tower were a major portion of the total heat load.

**Fig. 8.1** Detector tower with one crystal. Completed tower supports six crystals [17]

## 8.3 Cryogenic System General Description

Science requirements were for a cryostat with very low radioactive background, shielded against gamma and neutron radiation. Detector operation below the tungsten Tc requires multiple stages of heat shielding. The cryostat had to be large enough to contain seven detector towers. The low background requirement strictly limited materials that could be used.

The CDMS ICEBOX provided the mounting, thermal shielding and cooling for the CDMS detectors. The ICEBOX includes six nested cans with top access, a stem (E-stem) to carry out the signal cables, a stem (C-stem) for heat conduction, and aramid rope suspension. The inner can held 0.022 $m^3$ volume. The CDMS Icebox at Soudan operated for twelve years including multiple runs over one year in duration.

Figure 8.2 shows the overall layout of the experiment, showing the detector space surrounded by Icebox, a magnetic shield, lead and polyethylene shielding. The assembly was installed in an RF shielded, class 10,000 clean room 2300 ft. below the surface at Soudan Underground State Park [18] in Soudan, MN (Fig. 8.3).

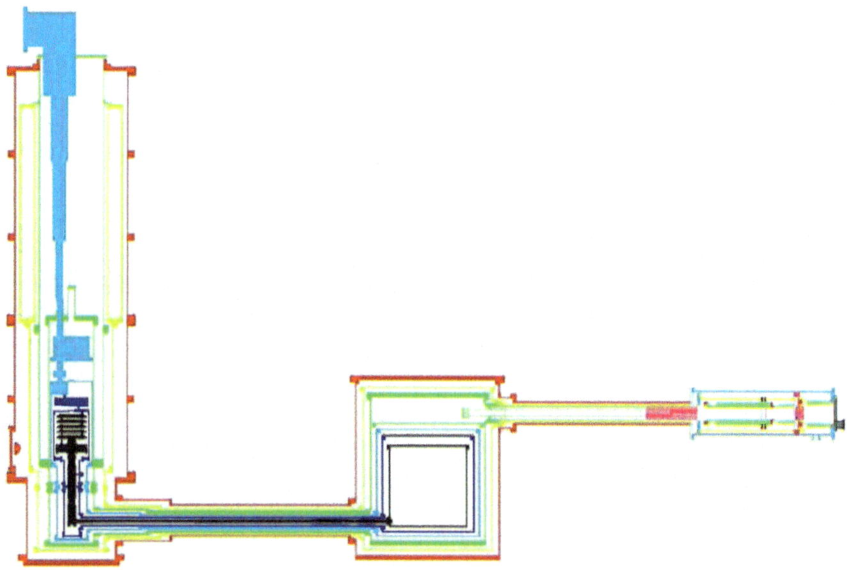

**Fig. 8.2** Overall cryogenic layout, dilution refrigerator on the *left*, ICEBOX cans in the *middle*. C-stem provides conductive cooling from the ICEBOX to the fridge. The signal feedthrough is on the *right* [17]

**Fig. 8.3** Icebox top, showing tops of the OVC, shield can, IVC can with five towers [17]

## 8.4 Dilution Refrigerator Introduction

The next sections refer to dilution refrigerator cooling. For the reader unfamiliar with this type of refrigerator the description of a 'wet' can be found in general literature, for example at [7].

## 8.5 Icebox General Description

The CDMS ICEBOX provided the mounting, thermal shielding and cooling for the CDMS detectors. The ICEBOX includes six nested cans with top access, a stem (E-stem) to carry out the signal cables, a stem (C-stem) for heat conduction, and aramid rope suspension. The stem length was needed to pass through the thicknesses of the lead and polyethylene shielding. Thermal models have confirmed that this number of layers is appropriate for the large milliKelvin detector operation. For CDMS the temperatures chosen were suited to the operating stages of the wet dilution refrigerator. Future large detectors may use dry fridges and may not be limited by helium or nitrogen boiling temperatures.

### *8.5.1 Icebox Cans*

Each of the cans provides heat shielding and mounting for the next inner can. Figure 8.4 shows the top view of the can assembly. The cans are referred to

**Fig. 8.4** ICEBOX cans, *top view*. The support adjustment screws for the outer layers can be seen at the *bottom* of the view. The signal cables exit through the E-stem on the *left* [3]

(moving from the outermost inward) as the Outer Vacuum Can (OVC), the Shield can (SH), the Inner Vacuum Can (IVC) can, the Still (ST) can, the Cold Plate (CP) can, and the Mixing Chamber (MC) can. The names refer to the associated dilution refrigerator stage.

The MC lid seen in Fig. 8.4 shows the seven hexagonal covers where towers can be installed. The purge tubes on the right are removed before closure.

The OVC was the outer vacuum enclosure. It was a vertical right cylinder 22.5 in. diameter and 27 in. tall. The walls were 1/8 in. thick and the top and bottom lids were 5/8 in. thick. The insulating vacuum is continuous with the outer vacuum of the dilution refrigerator. The removable lid and stem connections were sealed with Viton O-rings. The bottom of the OVC rested on the internal polyethylene shielding.

The Shield can operated at 80 K. It was wrapped with multilayer insulation (MLI) to reduce thermal radiation from the OVC. The inside was gold plated over nickel flash to reduce thermal emissivity. It is conduction cooled through the C-stem back to the liquid nitrogen reservoir in the dilution refrigerator. The Shield Can is not vacuum leak tight, but does provide a complete thermal radiation barrier and support heat intercept for the IVC. The shield can was 20 in. in diameter and 24.4 in. tall.

The IVC can operated at 5 K. The inside and outside were gold plated to reduce thermal emissivity. It was conduction cooled through the C-stem back to the liquid helium bath in the dilution refrigerator. The IVC separated the inner and outer vacuums. The inner vacuum was continuous with the dilution refrigerator inner vacuum and with the electronic feedthrough box. This can contained two lids; and inner lid that provided termination and final heat sink for the signal cables, and an outer lid that served as a vacuum tight cover.

All seals inside the background shielding were made with gaskets of annealed C101 copper sheets pressed between two bull nose raised surfaces with a 0.044 in. radius. Alignment pins on the outer lid and the stem connections ensured that opposing bull nose ridges matched each other. Fastening screws applied enough force to yield the annealed gaskets. Figure 8.5 shows the bull nose design. Seals outside the background shielding were made with indium wire. The IVC was designed to withstand full vacuum or atmospheric pressure with or without vacuum in the OVC.

The main purpose of the IVC was for detector cool down. The inner layers are thermally isolated from one another both in the cans and the dilution refrigerator, good for normal operation but not for cool down. By adding helium gas at a pressure of 0.1 atmosphere inside the IVC, inner cans and detectors become thermally well connected. This gas remains in place during cool down until a temperature of about 7 K is reached, at which time it is evacuated. The OVC remains fully evacuated during cool down, maintaining the MLI performance.

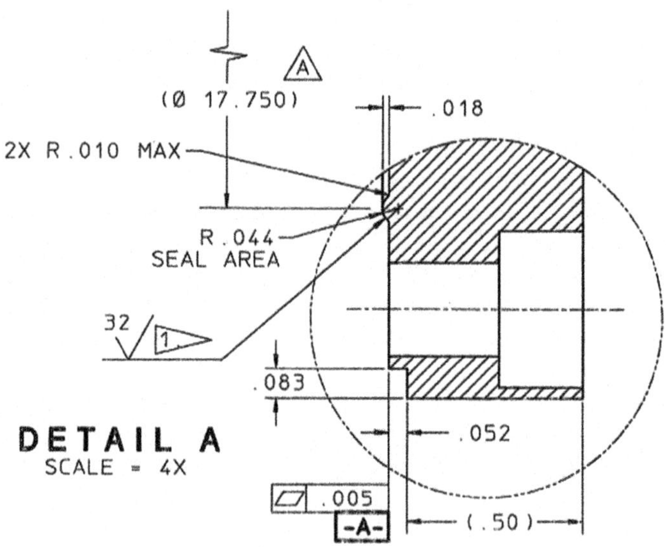


**Fig. 8.5** IVC seal design [11]

The Still can operated around 1 K. It was conduction cooled through the C-stem back to the dilution refrigerator's still stage. The Still can is not vacuum tight but does provide a thermal radiation barrier and heat intercept for the CP layer supports. The Still can is 16 in. diameter and 16 in. tall.

The CP can operated around 230 mK. It was conduction cooled through the C-stem back to the dilution refrigerator. The refrigerator connection was at a midpoint in the counterflow heat exchangers between the still and the mixing chamber. The CP can is not vacuum tight be does provide a heat intercept for the MC layer supports. The CP can is 14.3 in. diameter and 14.3 in. tall.

The MC is the innermost can and operated at 56 mK. It was conduction cooled through the C-stem back to the dilution refrigerator. This can was connected to the dilution refrigerator mixing chamber. It provided heat sinking for the detectors and supported the weight of the detectors and towers. Both of those functions were carried out through the lid, which was made of a flat plate with hexagonal openings. It was 12 in. diameter and 12 in. tall.

### 8.5.2 Suspension

The OVC can rested on the inner polyethylene shielding. The five inner cans were each suspended on aramid fiber strings looped over pulleys attached to the top of the next warmer can. Aramid ropes have very low thermal conductivity and low elasticity. The loops were attached at the top of the warmer can and the bottom of the colder can. The adjustment screws and rope arrangement for each can allowed height adjustment and leveling. The innermost MC can had three equally spaced suspension loops. A flexible cable in the C-stem allowed some motion between the fridge and the MC can. The next four inner loops were suspended on two loops diametrically opposed and perpendicular to the stem attachment. For these four cans the dilution refrigerator provided the third support point, balancing the cans. The side loops were carefully adjusted to match the tail flanges with the fridge bottom flanges.

### 8.5.3 C-Stems and Tails

The Fridge Tails and Cold Stem connected the dilution refrigerator with the ICEBOX. All six layers were continuous to the associated layers in the fridge, which provided all cooling.

Changing the stem direction from horizontal to vertical was a challenging task. This was accomplished with five nested tees and a bolted conduction rod connected to the bottom of the fridge. The fridge bottom connection is shown in Fig. 8.6.

**Fig. 8.6** Bottom of dilution refrigerator [17]

## 8.6 E-Stem

The E-stem carried the signal cables from the top of the towers at 5 K to the room temperature feedthrough box. It extended the OVC which terminated at the outer end of the E-stem. The LN shield layer extended though the E-stem and terminated with a stainless steel bellows at the outer end. A signal cable heat intercept at 90 K was located at the outer end of the stem. The LN layer in the E-stem was gold plated for low emissivity.

The IVC also extended through the E-stem. It carried the signal cables with a 5 K heat intercept near the can end. The IVC vacuum extends through the E-stem and into the feedthrough box. The E-stem inside diameter was 2 in., large enough to handle the maximum number of signal cables and their heat sinks at 90 and 5 K. The inner, narrow portion of the E-stem passed through the gamma and neutron shielding. The larger, outer portion contained the bellows isolating the inner and outer vacuums and temperature layers.

### *8.6.1 Thermal Contraction*

Aramid loops have a slight thermal expansion, but the copper cans shorten when cooled, effectively lowering each can relative to the OVC. And since the dilution refrigerator is internally suspended from the top, it raises the balancing supports when cold. Therefore the ICEBOX cans lower and tilt when cooled down. The nested stems fit very closely together and the heights were adjusted so that they would have the maximum clearance cold. This meant that some adjacent stems were touching at room temperature but separated during cool down. Flexible connections in the E-stem allowed the LN and IVC to bend neat the can slightly during cool down, and bellows at the outer end of the E-stem allowed contraction along the stems.

### 8.6.2 Materials, Radiopurity

The materials used in the ICEBOX were strictly controlled to reduce the risk of radioactive contamination. Many common materials used in cryostat fabrication contain small amounts of radioactive contaminants or isotopes, so consequently they could not be used. Copper alloy C101 was the primary cryostat material, the highest purity available commercially. Although higher purity copper can be made with lower oxygen content, it is not available in the required quantities. One of the fabrication goals was to minimize cosmogenic activation which occurs after electrostatic refining when the material is above ground. The material was purchased with the most recent refining date reasonably achievable. Then the copper was stored underground when not needed and brought to the surface for fabrication.

Fasteners were primarily socket head brass screws and brass alignment pins. These were custom made from free machining brass rods, again for material control. Temperature sensors were $RuO_2$ or platinum resistors, connected with phosphor bronze wiring, and either polyimide or Teflon wire insulation. Millmax pins and counted solder were used to terminate sensor wiring inside the ICEBOX.

Samples of all materials used were tested in a low background radiation counter before fabrication.

### 8.6.3 Fabrication

All machining was done with carefully cleaned work areas and new tools to reduce the risk of embedding foreign material. All Icebox welding was done with electron beam process so that no filler metal was needed. Following fabrication, the copper assemblies were cleaned in an acid bath to reduce radon daughters and other surface contaminants acquired during fabrication, rinsed with water and wrapped for cleanliness. Gold plating over nickel flash will be applied where needed for thermal emissivity. Following cleaning all handling was done with powder free clean room gloves. One enhancement that could have improved thermal conductance of the joints would have been gold plating of all joint surfaces.

Tapped holes were made with rolling taps. This reduces the risk of embedded tool particles from cutting taps. It has the additional benefit of work hardening the threads, increasing their strength.

The conductive tubes and rods in the C-stem were annealed to achieve high thermal conductivity. The annealing was held at 410 °C in vacuum for one hour and slowly cooled. This process followed a procedure described by Fickett [6] to achieve a high RRR in oxygen free copper (see Fig. 8.7).

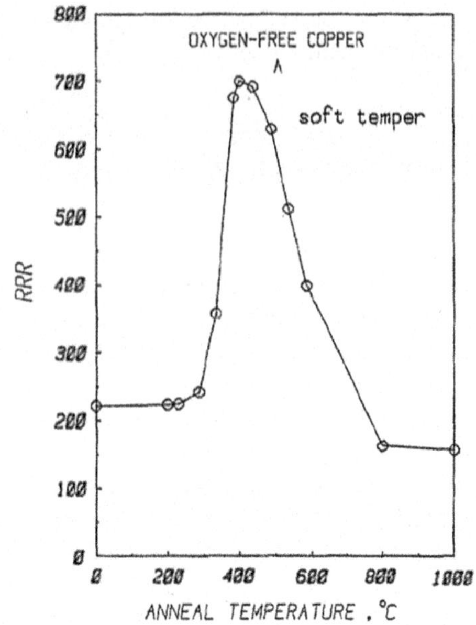


**Fig. 8.7** Anneal temperature for maximum RRR [6]

## 8.6.4 *Underground Assembly*

All parts were again wiped with alcohol before introduction into the clean room. All handling was done with using cleanroom gloves, lint free cleanroom wipes, cleaned fixtures and tools. The nesting of the cans, stems and dilution refrigerator attachment required a specific assembly sequence. Icebox cans and the associated C-stem and E-stem were assembled from the outside in. The OVC, outer vacuum container, was mounted and leveled on the lead shielding. The OVC C-stem and E-stem were attached, temporary covers applied and the joints were leak tested. All seals on the OVC were made with Viton O-rings.

The gold plated joints were cleaned with alcohol wipes. All bare copper joints were prepared for assembly by scrubbing the stem connections with counted Scotch Brite$^{TM}$ and cleaning with alcohol wipes. This removed surface corrosion immediately before assembly and the wipes removed any particles from the cleaning. The aramid loops were attached to the can and a lifting fixture was used to lower the LN can into the OVC can. The C-stem and E-stem were inserted into the OVC stems and connected from the inside. All screws were tightened following a pattern using a torque screwdriver. All screws were assembled dry and tightened to 90 % of their breaking torque. Tests had shown that the breaking torque was repeatable for both brass and stainless machine screws. And even stainless screws would break before damaging the rolled threads. Stainless screws were only used outside the lead and polyethylene shielding.

## 8.7 Thermal Model

Thermal modeling of detector and cryostat elements is crucial, but with the paucity of consistent low temperature data and joint conductance information can be a challenge. Material properties above 4 K are readily available from NIST Cryogenic Properties Database (NIST) and the older Brookhaven National Laboratory selected cryogenic data notebook [9].

### *8.7.1 Thermal Conductivity*

Conduction through solids below 4 K requires data from various sources. For CDMS the two materials of consequence were Kevlar (aramid fiber) and copper. Plotting test results from several reports shown in Fig. 8.8, the NIST (NIST) and Ventura [26] data appeared to be consistent over a wide temperature range. And with higher conductivity than other reports they are the appropriate choices for conservative design.

Copper thermal conductivity at low temperature is highly dependent on purity, heat treatment and operating temperature. Extremely high thermal conductivity can be obtained with 5 or 6 nines copper treated to remove oxygen and annealed. However this is not practical for the large conduction tubes needed for a cryostat such as the CDMS Icebox.

When large conductors were needed, the highest purity commercial alloy was used. C101, aka oxygen free copper, is readily available in industrial sized sheets, tubes and bars. Annealed copper following the procedure recommended by Fickett can achieve reasonably high conductivity (see Fig. 8.7) [6]. Risegari reported test data for annealed oxygen free copper between 30 and 150 mK. Risegari [19] Extrapolating the NIST data for RRR = 150 below 4 K, assuming a linear relationship with temp between 5 and 10 K does not line up very well with the Risegari equation.

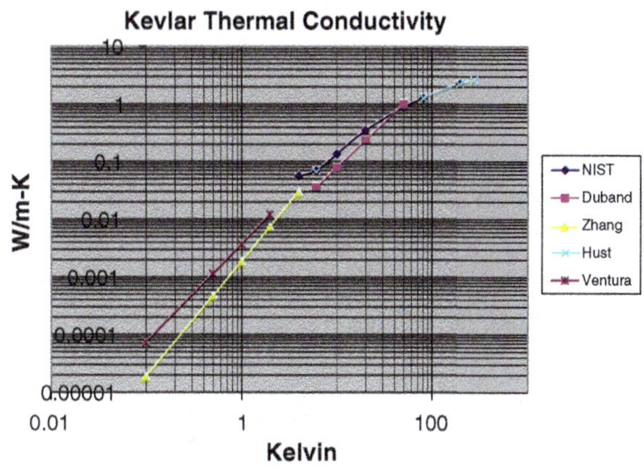


**Fig. 8.8** Low temperature thermal conductivity of aramid fibers (NIST) [5, 8, 12, 14, 26]

### 8.7.2 *Joint Conductance*

Bolted joints are required for assembly of most cryostats but can add a significant conductance resistance in the millikelvin range. Copper surfaces readily corrode, and even a small amount of corrosion can impose a significant decrease in joint conductance. For CDMS the major can and stem connections were rigorously scrubbed and cleaned immediately before assembly and performed reasonably well. But it early operation it was clear that joints assembled with less rigor performed poorly.

For SuperCDMS a study of over twenty-five publications regarding joint thermal conductance was performed. Unfortunately, a large majority of the results were not directly applicable to the SuperCDMS experiment because:

- Joint conductance results relied on electrical resistivity measurements and the use of the Wiedemann-Franz Law, whose use across joints was questioned in Nilles [13] and Didschuns [4]
- Joint clamping force was unknown
- Use of foreign materials such as grease between the joint surfaces. Use of interfacial materials is to be avoided in SuperCDMS to maintain experimental radiopurity and general cleanliness.

Upon review of the pertinent publications; Nilles [13], Kittel [10], Didschuns [4], and Woodcraft [29], a design basis was chosen. Parameters for surface finish, plating and clamping force were chosen that could be readily achieved in a project with dozens of conductive joints and hundreds of bolts. Subsequent tests from 60 mK to 14 K in three refrigerators confirmed the validity of this design equation (Eq. 8.1) [24]. The design basis and test results are shown in Fig. 8.9.

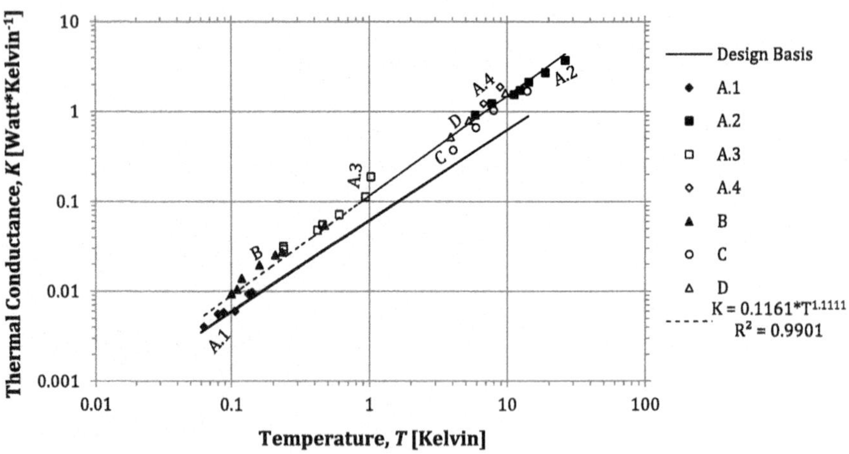


**Fig. 8.9** Gold plated copper joint conductance, design basis and test results [24]

$$K = 0.0624\,T - 0.00023 \tag{8.1}$$

where:
K Joint Conductance [W/K]
T Temperature [K]

## 8.8 Heat Load

Heat load calculations between stages were dominated by the detector tower structure, signal wiring and wiring insulation. The thermal conductivity varies widely as a function of temperature and a list of relevant papers for design below 1 K may be most useful for the readers

Nb-Ti, [15]
SS316, [1]
Kapton HN, [2]
Kapton HN, [20]
Graphite CFRP, [21]
Ti15333, [27]
POCO AXM-5Q, [28]
Manganin 4 % NI, [16]
Manganin 4 % Ni, [25].

## 8.9 Detector Signal Feedthrough

Vacuum feedthroughs are readily available for a variety of standard connector types. But when hundreds or thousands of signal wires must be accommodated the cost and space requirements of standard commercial feedthroughs can become prohibitive. A variety of designs have been used. CDMS constructed a feedthrough box from an ISO 400 six way cross. Two sides of the cross were fitted with vacuum pumping or cryostat connections. Eleven D50 connectors were welded into blank ISO 400 flanges on the other four sides. Strain relief was provided by brackets on the outside of the flanges.

Another approach is to press a printed circuit board between vacuum flanges. Connectors in the middle of the flange face inward. The signals traces can be single or multilayer and extend radially through the seal area of the board. Connectors outside the sealing surface face outward. Choice of seal type and number of board layers is mainly driven by space considerations.

## 8.10 Dilution Refrigerator

CDMS at Soudan used a wet (consumes liquid nitrogen and liquid helium cryogens) dilution fridge for cooling at all stages. Later modifications added cryocoolers for heat interception in the E-stem, nitrogen and helium re-condensation.

After startup difficulties were resolved, the dilution refrigerator was very reliable, with multiple runs over one year duration. The longest run was nineteen months, and these later runs were terminated for scheduling reasons, not fridge problems.

Several enhancements improved the reliability of the dilution refrigerator.

Larger, separate liquid nitrogen reservoirs replaced the original, single reservoir for the 80 K cold traps. This allowed temperature and level measurements to be added for automatic nitrogen filling. The original 80 K cold traps were used for the entire project.

The circulation pump with mechanical shaft seals was replaced with a magnetic drive pump, eliminating possible shaft seal leaks and seal maintenance. The pump oil was replaced with Fomblin SV for low oil vapor carry over. After draining and filling with the replacement oil, it was again drained and heated in a beaker, from which the original oil was skimmed off. The reliability of this pump has been excellent.

The dilution refrigerator was designed with a cold trap inserted in the bath. A second, external 4 K cold trap was added upstream and in series with the fridge internal cold trap. The external cold trap resided in a separate liquid helium bath. This external cold trap was initially filled manually from liquid helium supply dewars, but soon was equipped with a cryocooler helium liquefier, which maintained the liquid helium without further transfers. Very small amounts of helium leakage were made up with gas from high purity gas cylinders.

Regular, monthly cold trap regeneration was the key to long term reliable operation.

Most new experiments utilizing dilution refrigerators choose ‘dry’ models, which use a cryocooler(s) rather than liquids nitrogen and helium to provide the upper stage cooling. The dry fridges avoid the cost of liquid helium, the effort of dewar handling and liquid transfers. The latter is especially valuable underground or where access is limited.

## 8.11 Liquid Transfer Systems

The dilution refrigerator at Soudan was a wet dilution refrigerator of a type typically installed at the time. With the heat loads from the Icebox, it required daily transfers of liquids nitrogen and helium. Common practice for many refrigerators at that time was to manually transfer cryogenic liquids. But the Soudan underground location with limited access required an automated solution. It was also important to use liquid cryogens efficiently to minimize cost and effort to transport dewars underground.

Dual 160 L nitrogen supply dewars were connected to a vacuum insulated manifold outside the RF room. During transfers, liquid was drawn from the lead supply dewar. If it ran dry during the course of a transfer the backup dewar would complete the transfer. The backup would then become the lead for the next transfer, allowing time for the empty to be replaced. Inventory was tracked by scales located under each dewar.

160 L nitrogen dewars normally have pressure building regulators and vent valves. But those regulators often provide imprecise control and/or leak through wasting liquid nitrogen. More precise pressure control contributes to faster transfers without risking an overpressure in the fridge. When the liquid withdrawal line was connected to a dewar, a gas management line was also connected. The gas management line was equipped with pressure measurement, gas makeup and vent valves. During standby the PLC operated solenoid valve vented nitrogen if the pressure was too high. During transfers pressure was maintained with another solenoid valve supplied from high pressure cylinders.

When a transfer was initiated, a gas line pressurized the supply dewar and a cool down valve opened. When the transfer line was cooled down, the valve(s) filling the dilution refrigerator or 80 K cold traps opened and the cool down valve closed. A valve sequence after filling was used to ensure transfer efficiency and prevent trapped volumes in the piping.

Dual 350 L liquid helium dewars were connected to a vacuum insulated manifold outside the RF room. This was the largest helium dewar size that would fit on the mine shaft elevator. Dewar selection, pressurization, inventory and transfer procedures were similar to the nitrogen system. The helium transfer valves were close to the dilution refrigerator inside the RF room and therefore were pneumatically controlled from outside.

To minimize detector down time the helium and nitrogen transfers could be automatically synchronized. Transfers could be initiated by time-of-day, by low nitrogen or helium level in the fridge, or manually.

## 8.12 Liquefier Addition

The initial operation of the dilution refrigerator required daily transfers of both liquids. During transfers the physics data taking was stopped, losing an hour each day of experiment run time. Transporting dewars underground was typically done weekly, taking several hours with the elevator doors removed, technicians at the top and the bottom of the shaft. Normal deliveries to the remote site were weekly although special shipments could be made.

As the commercial development of cryocoolers progressed it became feasible to add liquefiers to the system. For this project all cryogenic equipment had to be installed outside the RF room, complicating the installation.

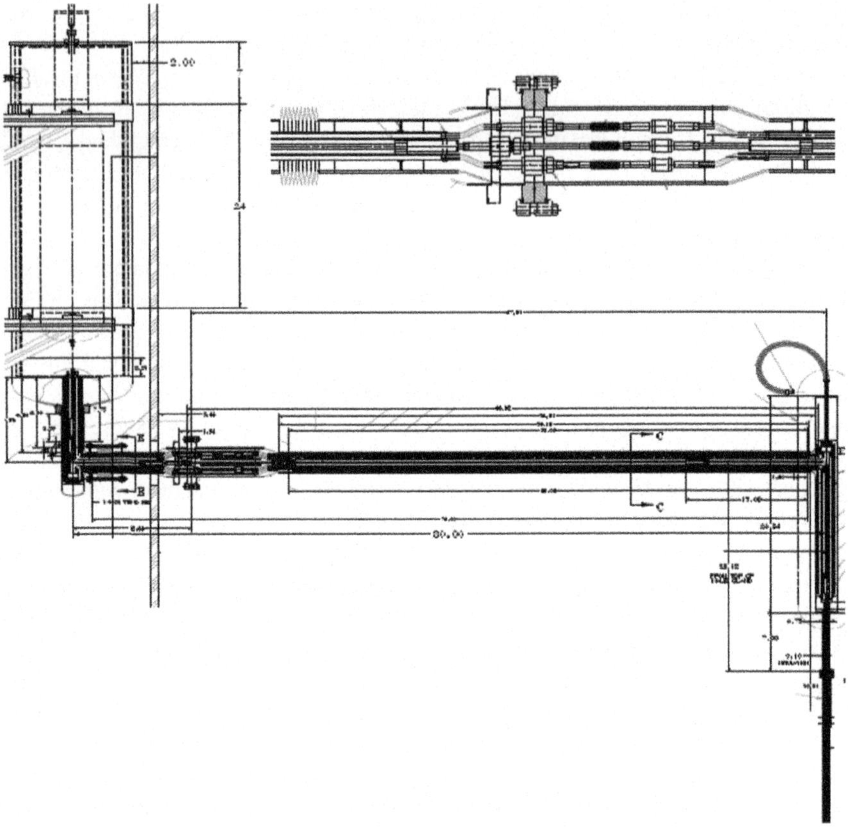


**Fig. 8.10** Low heat load transfer line between liquefiers and dilution refrigerator [23]

A transfer line was needed between the top of the dilution refrigerator to a bayonet box outside the RF room. The RF wall penetration, the thermo siphon design and the low heat load requirements prevented the use of a U-tube separation. Instead the transfer line had an S shape from the liquefier connection to the fridge. The line included an actively cooled 78 K heat shield, which extended into the male bayonet into the fridge. Assembly was made at the midpoint with a retractable bellows and VCR connections. A spare port on top of the fridge was modified to accept a custom male bayonet. The transfer line total length was 100 in. The transfer line is shown in Fig. 8.10.

The nitrogen liquefier was a GM type cryocooler was fitted with a desuperheating/condensing heat exchanger and installed in a custom vacuum vessel. An electric heater embedded in the heat exchanger was used to remove excess refrigeration capacity. The cryocooler vacuum vessel was mounted on rubber isolators to reduce the vibration forces carried into the support structure. During holding operation the heater was controlled by the nitrogen reservoir pressure.

Small losses were replenished by automatically adding gas from the liquid nitrogen supply dewars. The nitrogen liquefier system was very reliable, typically operating with about 120 W of excess capacity.

The dilution refrigerator was designed to use the helium boil off gas to intercept heat through the bath neck. Superheating helium gas also provided some cooling the nitrogen reservoir. Consequently adding a helium condenser would not have been feasible. The steady state heat load from the Icebox and dilution refrigerator consumed about 20 liquid liters/day. This was marginally within the capacity of two Cryomech PT-415 liquefiers, but with the transfer line heat load they were not quite enough capacity. With a third liquefier the cooling capacity was more than enough for the system. Electric heaters in the liquefiers were used to compensate for the excess cooling power. Heater power control was based on helium bath pressure.

Liquid helium from the refrigerator bath fed the 1 K pot. Helium from the pot was not recovered, so that loss plus leakage required regular makeup. Makeup gas was taken from helium supply dewar boil off. Although this system could run for months at a time, eventually trace amounts of contaminants would build up in the transfer line and stop the thermo siphon. Recovery involved disassembly of the transfer line, removal from the dilution fridge and a warmup. This could normally be done in one day without disturbing the fridge base temperature operation. A description of the liquefier system was published in Advances in Cryogenic Engineering [23]. The source of contamination was not known, but an easier means of removing and warming up the transfer line would have been an improvement.

## 8.13 External Cold Trap

Small amounts of contaminants in the mix would gradually plug the internal cold trap. To improve the operating reliability a 4 K external cold trap was installed downstream of the 80 K traps. This trap dewar was initially filled with liquid helium transfers, but was fitted with a helium liquefier. The liquefier could handle the total heat load and slowly fill the cold trap dewar from high pressure gas cylinders. The liquefier heater was controlled by dewar pressure, which could also add or vent helium gas when required. This system was very reliable.

## 8.14 E-Stem Cryocooler

During initial operation with fewer detectors it became apparent that the E-stem conduction and heat sinks would not be adequate for more detectors. A two stage SHI cryocooler was fitted to the E-stem to intercept signal cable heat. To minimize vibration it was independently suspended and equipped with vibration isolation. Extra thin, edge welded bellows isolated the vacuum shell from the ICEBOX. Braided cables isolated the first stage and bundled fine wires isolated the second stage. Even with this support and isolation system noise from the GM cryocooler was troublesome for the detectors.

## 8.15 Insulating Vacuum

The insulating vacuum inside the system must be good to reduce the heat load, and it is very important to limit the helium that could enter the IVC through leaks or permeation. Dry roughing and turbo molecular pumps were used to evacuate the ICEBOX. They were equipped with automatic valves that close in case of power loss or pump failure.

Standard dilution refrigerators have evacuation tubes sized for typical small applications. These tubes are not large enough to reasonably evacuate the larger volumes in the ICEBOX. To improve the pumping speed for the OVC an adaptor and custom radiation baffle was added to the bottom of the fridge tails. It was the equivalent of a 3 in. diameter pumping line and had no noticeable effect on the helium bath heat load. Pumping lines were added to the signal feedthrough box for the IVC.

## 8.16 Automation and Control

The cryogenic system was monitored and controlled with an industrial PLC (programmable logic controller) and HMI (human machine interface). They were both current industry products during installation and most of the operating period. The automation and control system was backup up by UPS (uninterruptable power supply) and standby generator.

This system provided local and remote monitoring, local and remote control, historical cryogenic data collection and fully automated operation. Automatic cryogen transfers, recovery from power outages, liquefier heater control and so on were fully automated.

The millikelvin systems at Soudan did not have temperature control, generally running as cold as possible. Helium and nitrogen liquefiers on systems on the other hand must be controlled to avoid subatmospheric pressure and freezing the nitrogen. Resistance heaters in the liquefiers, with zero crossing SCR's were controlled using conventional PID loops with cryostat pressure as the process variables. Pressure measurement is sensitive and small measurement errors will not cause the system to operate below atmospheric pressure.

Much more complicated are the sequences, abnormal procedures and alarms. Some examples from Soudan included conditional alarms, automatic set point adjustment, restart sequences after power outages, backup cooling water start, cryogen transfer sequences, etc. A modern industrial PLC with IEC 61131-3 programming languages can be configured to carry out these automation tasks in graphical form, without resorting to complicated scripts.

## 8.17 Cryogenic Operation

The remote location made commissioning and early operation more difficult. During the early years there were several operational problems that required the fridge to be removed from the ICEBOX. Once settled, however the fridge operation became very reliable. There were multiple runs at 56 mK over a year in length with the longest at 19 months. These runs were usually stopped for reasons unrelated to the dilution refrigerator. The dilution refrigerator and ICEBOX were in service for twelve years.

## 8.18 Lessons Learned

### *8.18.1 Cryogenic System Assembly and Testing*

After successful shop tests there was an assumption that after shipping and reassembly the dilution refrigerator would operate successfully. But several cool down attempts were required to commission the fridge and support equipment, repair leaks and damage, etc. Repairs were made more difficult by the underground, remote location. Future experiments in remote locations should consider a full operational test of all cryogenic systems at a convenient location.

### *8.18.2 Wiring*

Thermometer and touch sensor wiring was routed through the dilution refrigerator, which was equipped with cryogenic rated micro D connectors and heat sinks at each stage. While this was thermally and functionally successful, routing these cables along the narrow C-stem was difficult. A better solution would be to include these signal wires along with the detector signal cables.

### *8.18.3 Mixture Purification*

The standard fridge came equipped with a 4 K cold trap immersed in the helium bath. Anything that gets past this trap can cause plugging at the condenser impedance and requires a subsequent warmup. The installation of an additional 4 K cold trap reduces the risk of trap plugging during long runs.

### 8.18.4 Micro Vibrations

The initial installation did not utilize cryocoolers, but their installation caused noise pickup in the experiment detectors. Even with typical mechanical isolation methods the transmitted noise was significant. It is thought that when excited, the SNOBOX cans vibrated at their natural frequencies, then transmitted the vibration through the suspension to the adjacent inner cans. Transmissibility could have been reduced by installing springs on the hanger supports, especially at higher frequencies. Future projects should consider transmissibility of impulse from outside the cryostat or even the negative spring constant concept.

### 8.18.5 Can Supports

The aramid fiber loops effectively supported the cans with low heat load. However they were constructed of sixteen loops each. The inelasticity of Kevlar string required the loops to be identical in length and consequently a more difficult assembly procedure. More recent projects have used a single loop made of a larger braided rope, simplifying assembly.

### 8.18.6 Inner Vacuum, Yes or No?

CDMS II at Soudan has a sealed Inner Vacuum Chamber (IVC), a separate vacuum space from the Outer Vacuum system. During cool down, above 10 K, this IVC space is filled with a small amount of helium exchange gas to thermally link the inner cans to the IVC. While successful, this approach requires a leak-tight chamber at 5 K and the ability to evacuate helium gas at 10 K before proceeding below 10 K. For CDMS the cooldown was accomplished with several steps: Introduce liquid nitrogen into the helium bath and nitrogen shield reservoir, purge all liquid and evacuate the nitrogen from the helium bath, slowly introduce liquid helium, evacuate the exchange gas. Then start the dilution refrigerator.

Since a dry fridge does not have liquid nitrogen or helium reservoirs, and since the typical two stage cryocooler does not have enough capacity to reasonably cool down a large cryostat, it is more appropriate to dispense with the IVC and use cooling tubes fixed to each stage for cool down. For a system of this type heat exchangers attached to each stage would be linked in series using stainless steel bellows. Helium gas coolant is circulated through these heat exchangers and out to cryocoolers. This allows a large single stage cryocooler to start the cooldown at room temperature, and a two stage cryocooler can take over below 40 K. Once the layers are cold the helium gas will be evacuated from this cool down circuit. The long conduction path provided by the bellows shape and the poor thermal conductivity of stainless steel limits the heat load between layers to an acceptable value once the helium is evacuated form this circuit.

## References


1. M. Barruci, Measurement of thermal conductivity of the supports of CUORE cryostat. Cryogenics **48**, 166–168 (2008)
2. M. Barucci, Low temperature thermal conductivity of Kapton and Upilex. Cryogenics **40**, 145–147 (2000)
3. CDMS at Fermilab (2003). Retrieved from http://ppd.fnal.gov/experiments/cdms/. http://titus.stanford.edu/public/photos/cdms_icebox.jpg
4. I. Didschuns, Thermal conductance measurements of bolted copper to copper joints at sub-Kelvin temperatures. Cryogenics **44**, 293–299 (1992)
5. L. Duband, Thermal isolation of large loads at low temperature using Kevlar rope. Cryogenics, 643–647 (1993)
6. F. Fickett, Oxygen free copper at 4 K: resistance and magnetoresistance. IEEE Trans. Magn. **19**(3) (1993)
7. G. Frossati, Experimental techniques: methods for cooling below 300 mK. J. Low Temp. Phys. **87**(3/4) (1992)
8. J. Hust, Low-temperature thermal conductivity of two fibre-epoxy composites. Cryogenics, 126–128 (1975)
9. J. Jensen, *Brookhaven National Laboratory Selected Cryogenic Data Notebook* (1980)
10. P. Kittel, Thermal conductance of gold plated metallic contacts at liquid helium temperatures. Adv. Cryog. Eng. **37**, 241–248 (1992)
11. LBL. (1993). IVC Seal Design
12. E. Marquardt, *Cryogenic Material Properties Database*. 11th International Cryocooler Conference (Keystone, Co, 2000)
13. M. Nilles, Effects of oxidation and roughness on Cu contact resistance from 4 to 290 K. Adv. Cryog. Eng. **34** (1987)
14. NIST (n.d.), *NIST Materials Measurement Lab, Cryogenic Technologies Group, Material Properties*. Retrieved from http://cryogenics.nist.gov/MPropsMAY/materialproperties.htm
15. J. Olson, Thermal conductivity of some common cryostat materials between 0.05 and 2 K. Cryogenics **33** (1993)
16. I. Peroni, Thermal conductivity of manganin below 1 K. Nucl. Phys. B **78**, 573–575 (1999)
17. D. Bauer, PAC Meeting, March 30, (2007). Retrieved from https://www.fnal.gov/directorate/program_planning/March2007PACPublic/BauerPAC03_07.pdf
18. M.D. Resources, *Lake Vermilion-Soudan Underground Mine State Park* (2015). Retrieved from http://www.dnr.state.mn.us/state_parks/lake_vermilion_soudan/index.html
19. V. Risegari, *The Art of Cryogenics: Low Temperature Experimental Techniques* (Elsevier, Oxford, UK, 2008)
20. D. Rule, Thermal conductivity of polypyromellitimide film with alumina filler from 4.2 to 300 K. Cryogenics **36**, 283–290 (1996)
21. M. Runyan, Thermal conductivity of thermally-isolating polymeric and composite structural support materials between 0.3 and 4 K. Cryogenics **48**, 448–454 (2008)
22. J. Sander, *Fermilab Today* (2015). Retrieved from http://www.fnal.gov/pub/today/archive/archive_2015/today15-02-06.html

23. R. Schmitt, Application of cryocoolers to a vintage dilution refrigerator. Adv. Cryog. Eng. (2011)
24. R. Schmitt, Thermal conductance measurements of bolted copper joints for SuperCDMS. Cryogenics (2015)
25. Y. Touloukian, *Thermophysical Properties of Matter: Thermal Conductivity, Metallic Elements and Alloys*, vol. 1 (IFI/Plenum, New York, 1970)
26. G. Ventura, Low temperature conductivity of Kevlar. Crygenics, 489–491 (2000)
27. P. Wikus, The electrical resistivity and thermal conductivity of Ti 15V–3Cr–3Sn–3Al at cryogenic temperatures. Cryogenics **51**, 41–44 (2011)
28. A. Woodcraft, Thermal conductivity measurements of pitch-bonded graphite at millikelvin temperatures. Cryogenics **49**, 159–164 (2009a)
29. A. Woodcraft, Thermal design and performance of the SCUBA-2 instrument 1-K and mK systems. Cryogenics **49**, 504–513 (2009b)

# Chapter 9
# Cryogenic Transfer Lines

**Jaroslaw Fydrych**

**Abstract** Transfer lines are common in cryogenic systems and are a form of cryostat. This chapter describes the requirements of transfer lines, surveys existing transfer lines and discusses issues such as modularization, routing, supports, thermal contraction, piping arrangement, materials, manufacturing and installation. The chapter concludes with a detailed case study of the design, manufacturing and performance of the XFEL/AMTF transfer line.

## 9.1 Introduction

Cryogenic transfer lines are typical components of almost all cryogenic systems. They are intended for transferring cryogenic fluids between two cryogenic devices [1]. Since the value of the cryogenic fluids is essentially in the thermodynamic states of their molecules, the transferring should not cause significant changes in the thermodynamic states of the transferred cryogens. It means that either the temperature increase or, in case of liquids, the vapor quality and also pressure changes should be negligibly small.

The simplest cryogenic transfer line is a vacuum jacketed pipe connecting two nitrogen dewars as shown in Fig. 9.1. In this example, the line is used for transferring liquid nitrogen (LIN). If the distance between the two dewars is short, the flowing nitrogen stays in the line for a short period of time. Then, the flowing nitrogen absorbs little heat and only a tiny portion of the nitrogen evaporates during the flow and the rest keeps its initial thermodynamic state. The relatively high value of the latent heat of nitrogen is obviously an advantage. However, if the line is very long the nitrogen stays in the line much longer. Then the flowing nitrogen absorbs much more heat. As a result, the vapor fraction of the nitrogen reaching dewar 2 is much larger. The inflowing nitrogen vapor obviously does not stay in the dewar but flows out; it counts as losses. The higher hydraulic resistance of the longer process line leads to a

---

J. Fydrych (✉)
European Spallation Source ERIC, P.O. Box 176, 22100 Lund, Sweden
e-mail: Jaroslaw.Fydrych@esss.se



J.G. Weisend II (ed.), *Cryostat Design*, International Cryogenics Monograph Series,
DOI 10.1007/978-3-319-31150-0_9

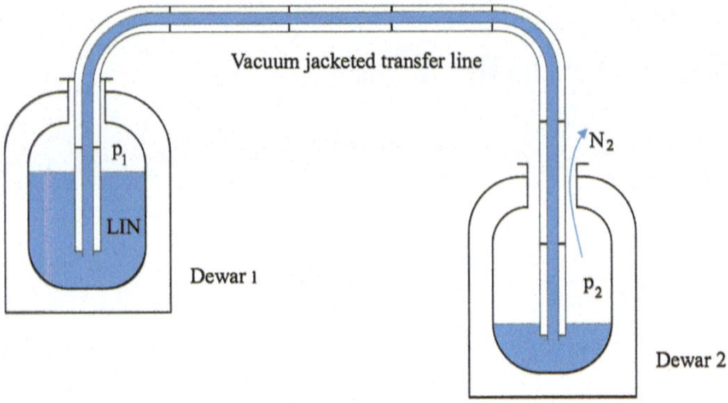


**Fig. 9.1** A simple pipe-in-pipe cryogenic transfer line connecting two dewars

higher difference between pressure $p_2$ and $p_1$, which increases the difference between the initial and final thermodynamic states in the transferred nitrogen.

The pressure, temperature and density of the flowing cryogen have certain impacts on the mechanical design of the process line, vacuum jacket envelope and their supports. The higher the pressure and density, the thicker the pipe walls and the stronger the supports.

This simple example of a pipe-in-pipe cryogenic transfer line shows that designers of these lines have to take into consideration a number of parameters. The most important are:

- geometrical restrictions (total distance, possible routings and available space),
- required mass flow rate and its time characteristic,
- minimum and maximum temperature,
- minimum and maximum pressure,
- thermodynamic properties of chosen cryogenic fluid,
- mechanical properties of chosen pipe materials.

The design of pipe-in-pipe cryogenic transfer lines is extensively described in [2–4]. All the above considerations and a number of design guidelines provided in [4] are also valid in the process of designing much more complicated cryogenic transfer lines, in which the vacuum jacket houses several process lines transferring cryogenic fluids at different temperatures and pressures. Such multichannel lines are usually used for transferring cooling power between a cryogenic plant and cryogenic users in so-called large scientific facilities. Typical cryogenic users are cryomodules with superconducting cavities and cryostats with superconducting magnets, bus bars or cryogenic vacuum pumps. These facilities usually require complex cryogenic systems that are capable of providing very high cooling power at very low temperature levels and distribute this cooling power among the users. Figure 9.2 shows a simplified schematic flow scheme of such a complex cryogenic system. Here, a cryogenic device, let's say a chain of cryomodules housing superconducting cavities made of pure niobium, needs to be cooled down to a temperature of 2 K. The required cooling power is produced by a cryogenic plant

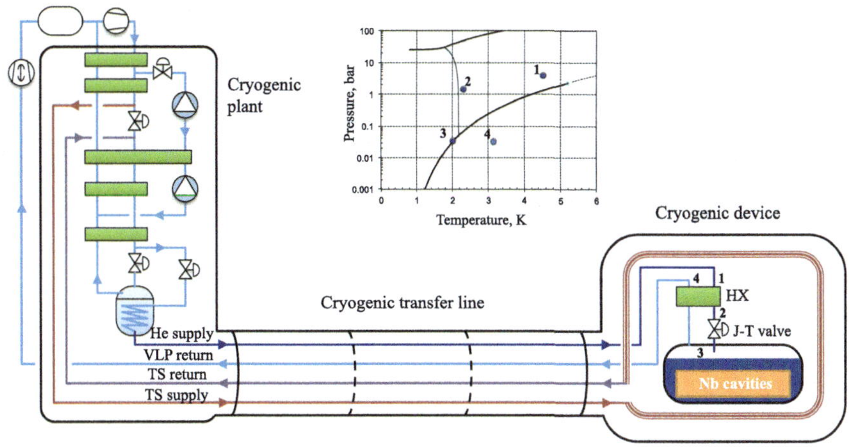


**Fig. 9.2** Schematic of a complex cryogenic system

and transferred by means of a constant flow of cold helium to the device via the helium supply line. In this example, the cryogenic plant supplies supercritical helium at a temperature and pressure of 4.5 K and 3 bar absolute, respectively. Then the 2 K helium is produced at the cryogenic device, where the supercritical helium is first precooled to 2.2 K in a counterflow heat exchanger (HX) and then throttled in a Joule-Thomson valve down to a pressure of 31.3 mbar absolute.

This solution takes advantage of a significantly low value of the helium critical pressure (2.25 bar absolute). Transferring helium at higher pressure eliminates all the problems related to two-phase flow phenomena. Usually a pressure around 3 bar absolute is perfectly adequate. Then, the cryoplant compressors have to compress helium to a pressure not higher then 20 bar absolute, which is not too complicated from technical point of view.

Since helium is a very expensive cryogen it must be recovered and recirculated to the cryogenic system. Therefore in the above example the helium vapor is recovered from the cryogenic device and transferred back to the cryogenic plant via the vapor low-pressure line (VLP). The vapor flow is driven by a vacuum pump at the cryogenic plant, which generates the subatmospheric pressure required for reaching 2 K in the helium vessel inside the cryogenic device. In the case of very large systems, which require significantly high cold helium flows, vacuum pumps are replaced by sets of cold and warm compressors.

All the cold elements of the cryogenic device are surrounded by a thermal shield. The shield is actively cooled by an additional helium circuit at a temperature of 40–60 K. The thermal shield circuit is composed of the TS supply and return lines connected to the high pressure helium line in the cryoplant cold box, downstream the first set of its heat exchangers.

In the above example the cryogenic transfer line is one of the main components of the cryogenic system. Apart from all the four process lines (He supply line, VLP return line, TS supply line and TS return line) it includes also thermal shields, vacuum jackets, vacuum barriers, supporting structures, thermal contraction

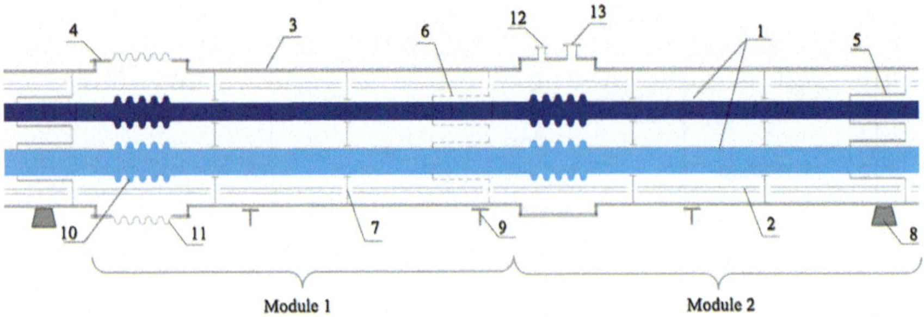


**Fig. 9.3** Schematic illustration of a multichannel cryogenic line

compensation elements, pressure safety devices, vacuum pumping ports and some instrumentation (temperature sensors and pressure transducers).

The cryogenic transfer lines are almost always manufactured and preassembled at manufacturer sites that are located far away from their final locations. Therefore, the lines have a modular structure that facilitates production, transportation and installation, as well as some repair work, if needed. This strongly affects design choices such as a number and lengths of sections (modules) and interconnection arrangements. Figure 9.3 shows a schematic illustration of a straight section of a multichannel cryogenic transfer line.

This section is composed of two straight modules. Each of them consists of process pipes (1), thermal shields (2) and an external envelope (3). The two sections of the external envelope are connected by an interconnecting sleeve (4). The cryoline section ends with vacuum barriers (5) that separate the cryoline insulation vacuum from those of the adjacent vacuum sectors. All forces resulting from the dead weights, pressure loads and thermal contractions of the process lines are transferred to the vacuum jacket via a process line fixed support (6), sliding supports (7) and the vacuum barriers as well, which also act as fixed supports. The external envelope transfers these forces to conventional facility structures (anchors or foundations at the building floor and walls) via the fixed supports (8) and sliding supports (9) of the external envelope.

Both the process pipes and the external envelope are equipped with bellows. Internal bellows (10) compensate the thermal shrinkage of process pipes, whilst the external bellows (11) are to cope with all the thermal expansions and contractions of the vacuum jacket. The external envelope is also equipped with a vacuum pumping port (12) and pressure relief device (13).

The most well known large scientific facilities that use multichannel cryogenic lines are listed in Table 9.1. In almost all these facilities, the cryogenic transfer lines are part of cryogenic distribution systems, which in addition to cryogenic transfer lines include a number of distribution boxes. For example the LHC cryogenic distributions system, which is the largest system in the world, is composed of eight separated cryogenic distribution lines. Each distribution line is 3.1–3.3 km long and consists of a number of cryogenic transfer line sections that connect 38 service modules (distribution boxes) [5, 6]. Another example is the RHIC cryogenic distribution system that includes 12 valve boxes and 25 sections of multichannel cryogenic transfer lines [7].

**Table 9.1** Examples of large scientific facilities using multichannel cryogenic transfer lines

| No. | Scientific facility | Institution (location) | Machine type | Cryogenic devices (operating temperature) | Cryogenic transfer line | | |
|---|---|---|---|---|---|---|---|
| | | | | | Total length | Cold line no. | Temperatures (cryogen) |
| 1 | ESS[a] | European Spallation Source ERIC (Lund, Sweden) | Proton linac and neutron source | RF cavities (2 K) | 450 m | 4 | 4–53 K (helium) |
| 2 | FAIR/SIS100[a] | GSI (Darmstadt, Germany) | Heavy-ion storage ring | Magnets (4.3 K) | 900 m | 4 or 5 | 4.3–100 K (helium) |
| 3 | FAIR/ SuperFRS[a] | GSI (Darmstadt, Germany) | Fragment separator | Magnets (4.3 K) | 500 m | 4 | 4.5–100 K (helium) |
| 4 | FLASH | DESY (Hamburg, Germany) | Electron linac with free electron laser | RF cavities (2 K) | 200 m | 4 | 4.5–80 K (helium) |
| 5 | FRIB[a] | MSU (Michigan, USA) | Proton linac and fragment separator | RF cavities (2 K) | 500 m | 5 | 4–55 K (helium) |
| 6 | HERA | DESY (Hamburg, Germany) | Hadron-electron storage ring and collider | Magnets and RF cavities (4 K) | 6.3 km | 4 | 3.7–80 K (helium and nitrogen) |
| 7 | ISR | CERN (Genewa, Switzerland) | Hadron storage ring and collider | Magnets (4.2 K) | 400 m | 2 | 4.2–100 K (helium) |
| 8 | ITER[a] | ITER Organization (Cadarache, France) | Tokamak | Tokamak magnets (4.3K), divertor cryopumps (4.5 K) | 3.5 km | 4–6 | 4.3–100 K (helium) |
| 9 | JT60SA[a] | JAEA (Naka, Japan) | Tokamak | Tokamak magnets (4.4K), divertor cryopumps (3.7 K) | 100 m | 5 or 6 | 4.4–100 K (helium) |
| 10 | KEKB | KEK (Tsukuba, Japan) | Electron-positron storage ring and collider | Crab cavities (2 K) | 1.1 km | 4 | 4.5–80 K (helium and nitrogen) |

(continued)

**Table 9.1** (continued)

| No. | Scientific facility | Institution (location) | Machine type | Cryogenic devices (operating temperature) | Cryogenic transfer line | | |
|---|---|---|---|---|---|---|---|
| | | | | | Total length | Cold line no. | Temperatures (cryogen) |
| 11 | KATRIN | TLK (Karlsruhe, Germany) | Tritium-neutrino experiment | Magnets (4.5 K) | 40 m | 6 | 4.5–117 K (helium and nitrogen) |
| 12 | LHC | CERN (Genewa, Switzerland) | Hadron storage ring and collider | Magnets (1.9 K) | 26 km | 4 and 5 | 1.8–80 K (helium) |
| 13 | RHIC | BNL (Brookhaven, USA) | Heavy-ion storage ring and collider | Magnets (4.6 K) | 5.2 km | 5 | 4.6–70 K (helium) |
| 14 | SNS | ORNL (Oak Ridge, USA) | Proton linac and neutron source | RF cavities (2.1 K) | 600 m | 2 | 2.1–50 K (helium) |
| 15 | Tevatron | FERMILAB (Batavia, USA) | Proton storage ring and collider | Magnets (5 K) | 6.7 km | 2 | 4.6–80 K (helium and nitrogen) |
| 16 | TORE SUPRA | CEA (Cadarache, France) | Tokamak | Tokamak magnets (1.8 K) | 100 m | 8 | 1.7 K–80 K (helium and nitrogen) |
| 17 | TRISTAN | KEK (Tsukuba, Japan) | Electron-positron storage ring and collider | RF cavities and magnets (4.2 K) | 330 m | 4 | 4.5–80 K (helium and nitrogen) |
| 18 | XFEL/AMTF | DESY (Hamburg, Germany) | Cryomodule test stand | RF cavities (2 K) | 170 m | 4 | 4.5–80 K (helium) |
| 19 | XFEL/LINAC[a] | DESY (Hamburg, Germany) | Electron linac with free electron laser | RF cavities (2 K) | 340 m | 6 and 7 | 2–80 K (helium) |

[a]under design or construction in Feb. 2016

## 9.2 Cryoline Routing and Modularization

The routing of a cryogenic transfer line defines its detailed location in the scientific facility site. In order to minimize both capital and operational costs the routing should be simple and the total length of the line should be as short as possible. However, in a great extent it is the geometrical features of the site infrastructure that determines space available for the line. An example of the most complicated cryoline routings is the ITER cryogenic distribution system [8, 9], which 3 D model representation is shown in Fig. 1.25 (Chap. 1). This system is to be located around the machine in the tokamak building.

Figure 9.4 shows a 3D model of the ESS cryogenic distribution line which runs from the cold box of the ESS accelerator cryogenic plant to the ESS linac tunnel in an underground gallery [10]. Since the gallery has chicanes, which are required for limiting the propagation of radiation from the tunnel to the cold box building, the transfer line is significantly elongated and has additional elbows.

Due to a complex internal design requiring a lot of precise assembly works the cryogenic transfer lines are usually produced in manufacturer workshops, which are located far away from the scientific facilities. In order to facilitate production, transportation and installation of cryogenic transfer lines, the lines are designed to be composed of a defined number of modules [11]. Each module is a short section of the cryoline that is connected to its adjacent sections via special interconnections. Producing and assembling cryoline modules at manufacturer sites reduces production costs and helps to meet high standards of production works. The prefabricated modules are transported to the site and linked together with the interconnections. The lengths and shapes of modules depend mainly on cryoline routings, transportation method and fixed support locations. Usually the lengths of modules follow the single

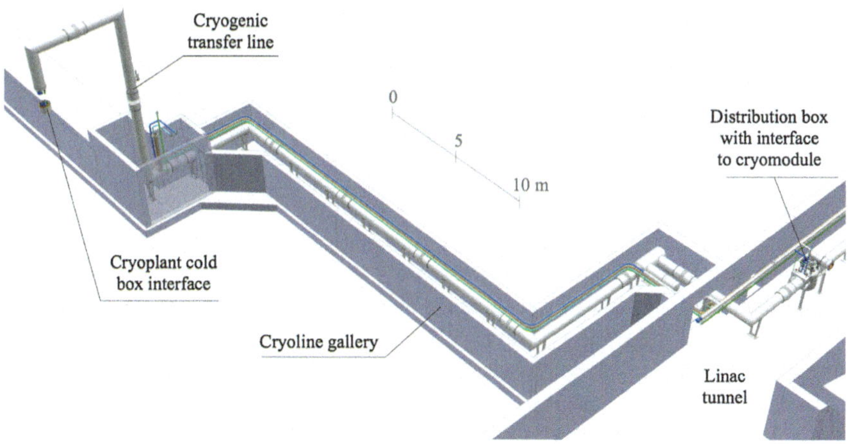


**Fig. 9.4** Schematic view of the ESS linac cryogenic transfer line

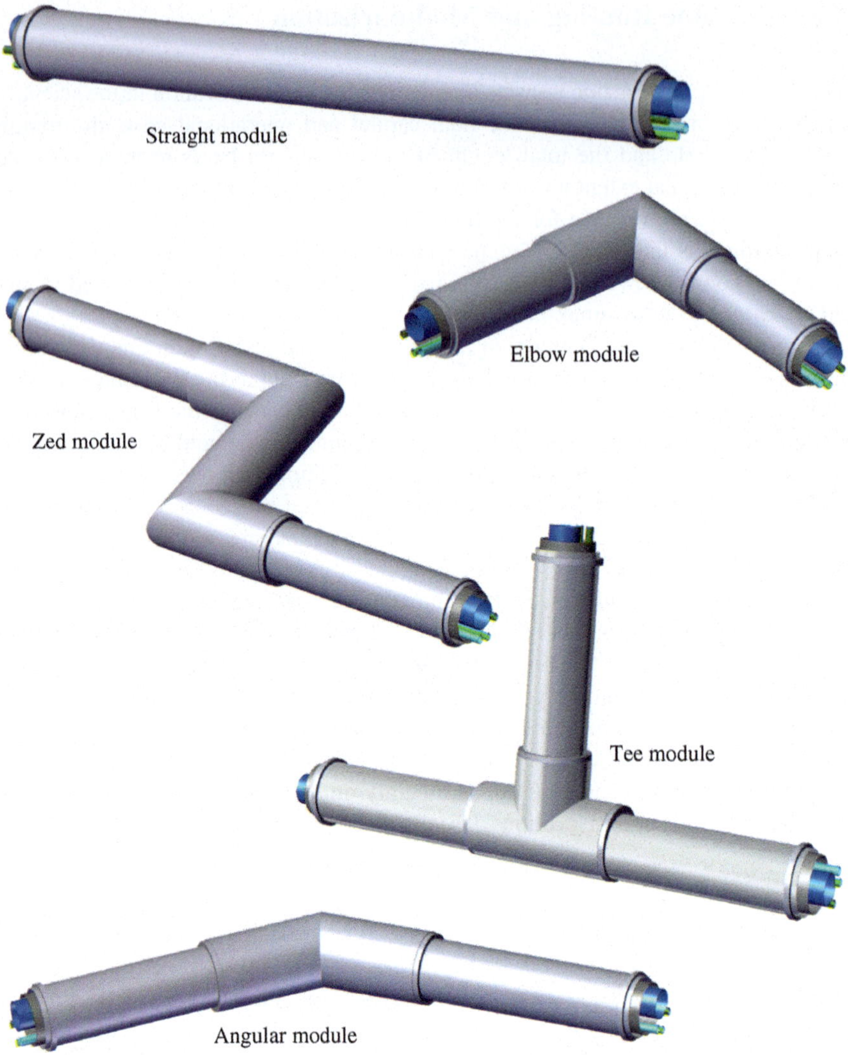


**Fig. 9.5** Typical cryoline modules

or double random lengths of pipes (the most commonly stocked lengths) and are not longer than the length of a typical semi-trailer (ca. 13 m).

Figure 9.5 shows some typical cryoline modules. The most frequently used are straight and elbow modules. The Zed module is used where the cryoline routing has a significantly small step, while Tee module is dedicated for distributing cryogenic fluids to two cryogenic devices (or two groups of devices).

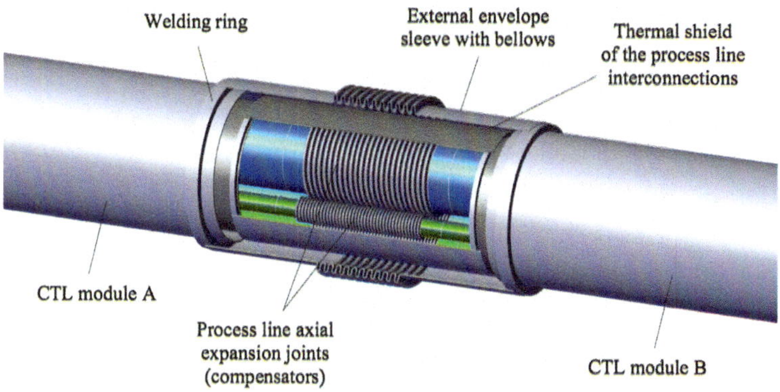


**Fig. 9.6** Typical design of an interconnection between two cryoline modules

During the installation of a cryogenic transfer line at its final location the cryoline modules are connected to each other with so-called cryoline interconnections. Figure 9.6 shows an example design of a cryoline interconnection.

The external envelope sections of two adjacent cryoline modules are ended with a welding ring that the interconnection sleeve is welded to. The sleeve is made of a pipe section which is slightly higher in size than the cryoline module external envelope. Such a sleeve can be moved aside on one of the modules in order to get an access to the internal parts of the interconnection. Some of interconnection sleeves are equipped with bellows allowing for required axial or lateral displacements. In this case, the internal process lines should also be equipped with axial or lateral compensators.

## 9.3 Cryoline Cross-Section Arrangements

The arrangement of process lines in a cryoline cross section can impact the thermo-mechanical behavior and thermal feature of the cryogenic transfer line. Therefore the detailed locations of process lines and thermal shield inside the vacuum jacket (external envelope) should be defined at one of the earliest design stages [11]. Proper arrangement of the lines should help to:

- minimize the external envelope size,
- minimize heat fluxes among process lines,
- reduce insulation vacuum space,
- avoid any unwanted thermal bridges among internal components,
- link thermally the thermal shield components with their cooling line,
- provide space for the supporting and thermal shrinkage compensation system elements (supports, spacers, bellows, metal hoses, etc.),

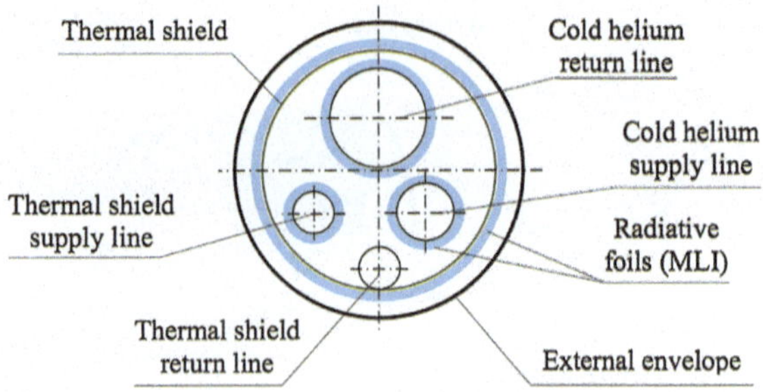


**Fig. 9.7** Example cross section of a four channel cryogenic transfer line

- keep enough room required for assembly (welding process lines and wrapping radiation foils in interconnections, etc.),
- keep balance in the distribution of process pipe and radiation shield dead weight to avoid significant moments which can lead to unwanted torsions of process pipes along the line.

Figure 9.7 shows a schematic arrangement of the cross section of a four-channel cryogenic transfer line. Its all process lines are surrounded by a cylindrical thermal shield wrapped with radiative foils (MLI). The thermal shield is connected with the thermal shield return line, which works as a thermal sink for radiative heat loads absorbed by the shield. Since the cold helium supply line has larger diameter than that of the thermal shield supply line, there is a clear asymmetry in the locations of these process lines in respect to the vertical center plane. This geometrical asymmetry guarantees the symmetrical distribution of process line dead weights.

Figure 9.8 collects the cross-sections of cryogenic transfer lines of some well-known scientific facilities. The cryogenic transfer lines of HERA, KEKB, LHC, Tore Supra, TEVATRON, TRISTAN, RHIC and XFEL/AMTF were already built and operated, whereas the cryolines of ESS, FRIB, ITER and XFEL/Linac injector are currently being designed or constructed. The TEVATRON CTL is a single path cryoline, what means that it is designed to transfer cryogenic fluids into one direction only. In order to return helium back to the cryogenic plant the cryogenic system needs to use two such lines (feed and return lines) or return the helium via a warm process line. All the other cryogenic lines shown in Fig. 9.8 are dual path lines. They consist of supply and return lines. The number of lines depends on the related cryogenic device cooling process and varies from four (eg. TRISTAN and HERA CTLs) to ten (Tore Supra CTL). The sizes of the process lines differ very much from line to line, since they depend on cryogen mass flow rate, density and allowable pressure drop. Process lines used for transferring cold helium vapour at low pressure usually have the largest sizes. In case of the LHC/QRL CTL this line is DN250 in size and it transfers helium vapour at a subatmospheric pressure of 16 mbar absolute and temperature of 4 K in a distance of 3.4 km. The diameters of the cryoline external envelopes also vary from line to line. The largest is the LHC/QRT CTL which is equal to 640 mm, while the smallest is the CTL of KEKB (156 mm).

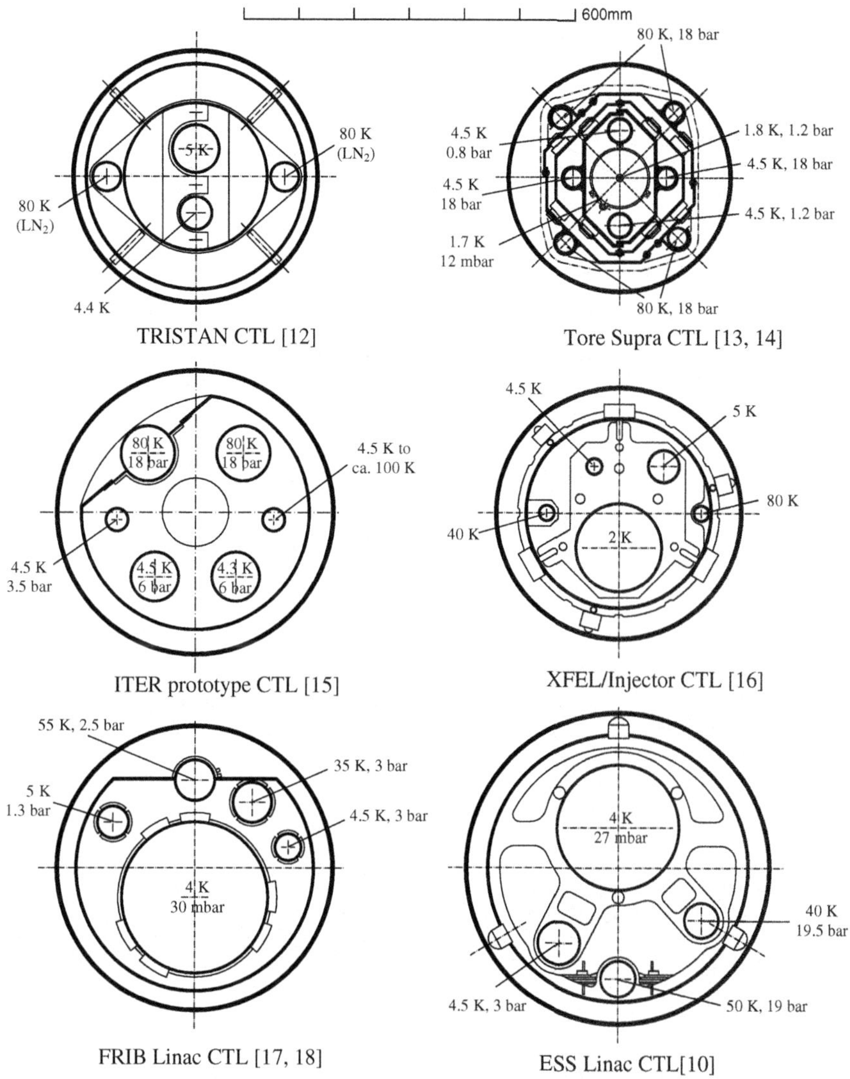


**Fig. 9.8** Cross-section arrangements of cryogenic transfer lines existing or under construction

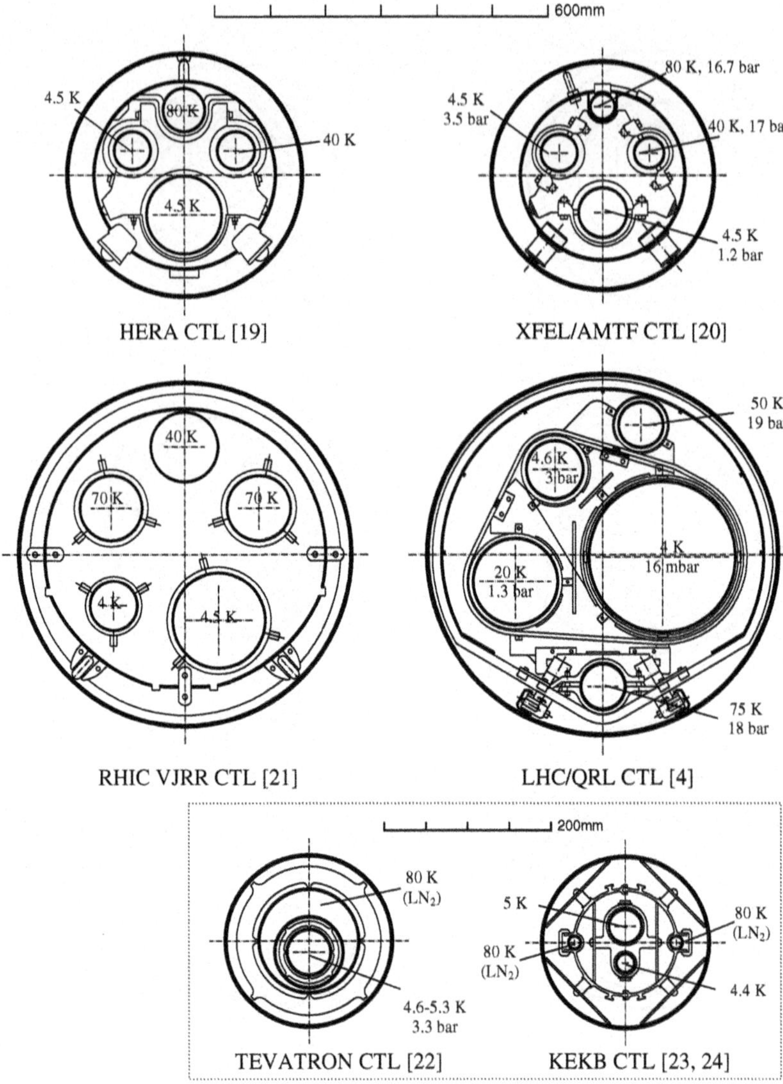


**Fig. 9.8** (continued)

## 9.4 Supporting Structures

Due to high temperature and pressure differences, and significantly high dead weights of cryogenic transfer line components, cryogenic transfer lines require complex supporting elements to carry the weights, to direct and transfer static and dynamic loads, as well as to control the movements of cryoline components. Forces resulting from pressure, thermal and gravity loads acting on the process lines (and in some cases on the thermal shields) have to be transferred to the external envelope and further to the surrounding structures, such as building floors and walls, piping bridges, etc. These forces can be in an order of magnitude of 10 kN or even higher. For example a DN250 process line equipped with an axial expansion joint (of an effective cross section $A_{ef} = 690\ \mathrm{cm}^2$), which maximum allowable pressure PS is 6 bar, at a test pressure of $\mathrm{PT} = 1.43 \cdot \mathrm{PS} = 8.58$ bar acts on its fixed support with an longitudinal force $F_x = A_{ef} \cdot \mathrm{PT} = 59.2$ kN. For a cryogenic transfer line that in addition to the DN250 process line includes a DN65 cold helium line (PS = 6 bar) and two DN50 thermal shield lines (PS = 25 bar) the total force that has to be transferred by the fixed support can reach 90 kN.

The supporting system of any cryogenic transfer line is composed of internal and external supporting structures. The internal supporting structures include fixed and sliding supports and spacers of process pipes and thermal shield components. Since these supports connect mechanically the cryoline components at highly different temperatures they should be designed in a way that significantly minimizes all unwanted heat transfers. For given temperature differences, Fourier's law of thermal conduction gives only three general ways for reducing heat flows: (1) elongating the heat transfer distances, (2) decreasing the heat transfer cross sections and (3) choosing materials with low heat transfer coefficients (see Chap. 1). All of these methods are applied in designing process line supports, however, the mechanical strength requirement limits the elongations, cross section decreases and applicable materials.

Process line fixed supports are supporting elements that bind mechanically the process lines to the external envelope and remove all six degrees of freedom of the lines in respect to the vacuum jacket. These supports are usually made of stainless steel in order to allow for welding their elements both to the process lines and vacuum jacket. Figure 9.9 shows a typical design of a process line fixed support. This support is composed of several welding rings and sleeves and a plate. Ring 1 is for welding the fixed support to the inner surface of the vacuum jacket. At working conditions the heat from the external environment flows via the vacuum jacket wall, ring 1 and sleeve 1 to the fixed support plate. The TS return line is thermally connected to the plate and works as a thermal sink that stabilizes thermally the plate at a thermal shield temperature (usually between 50 and 80 K).

The cold process lines and the TS supply line are connected to the plate via a set of two sleeves and two rings which elongate the heat flow paths. In the case of the cold helium return line, heat flows from the plate via sleeve 2, ring 2, sleeve 3, ring 3 to the wall of the line. Process lines can be also connected to the plate with one

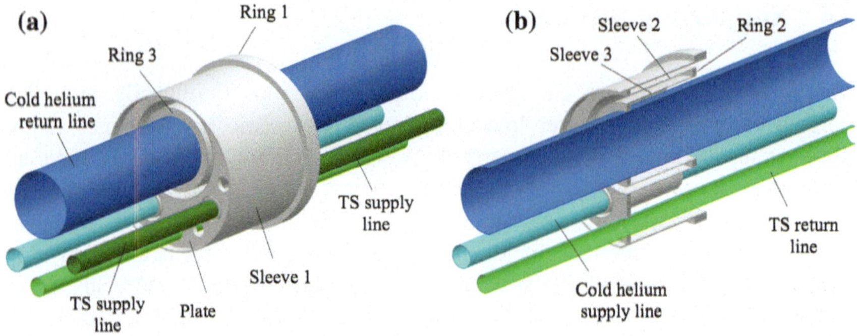


**Fig. 9.9** Process line fixed support: isometric view (**a**) and its longitudinal cross section showing elongation sleeves of cold process lines (**b**)

sleeve and one ring. Then the heat transfer path is shorter and more heat flows to the transferred helium, but the fixed support has better mechanical properties, and since its design is less complicated, its manufacturing is easier and less expensive. In order to reduce heat leaks the cryoline designers can elongate the sleeves and make them from tubes of thinner walls as long as the fixed support has appropriate mechanical parameters. In case of long sleeves there is a risk of lateral deformations of the sleeves when exposed to radial forces. It can lead to unwanted thermal bridges which can significantly increase heat loads to the cold process lines. The lengths of sleeves usually varies from 0.3 to 0.8 m, whist their wall thicknesses are usually similar to the process line wall thicknesses (1.8–3.2 mm).

Process line sliding supports are supporting elements that direct pressure and thermal loads to the fixed supports. They also carry the weight of a section of process lines, equipment attached to the lines (thermal shield components, radiation foils, getters, absorbers, instrumentation) and cryogenic fluids in the lines as well. Since these support do not bear high mechanical loads they are typically made of composite materials characterized by low thermal conduction coefficient. Figure 9.10 shows an example design of a process line sliding support. This support is composed

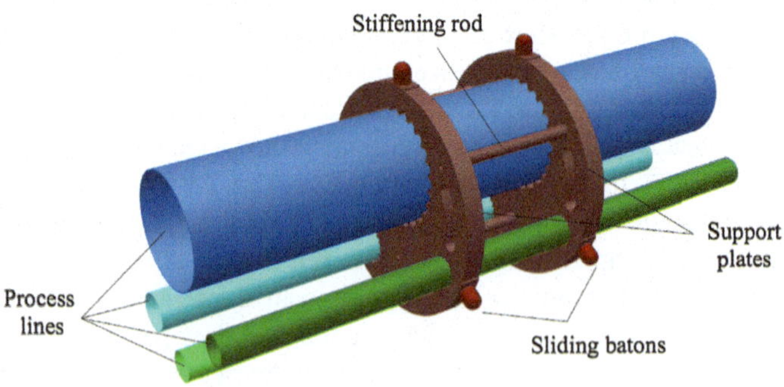


**Fig. 9.10** Process line sliding support

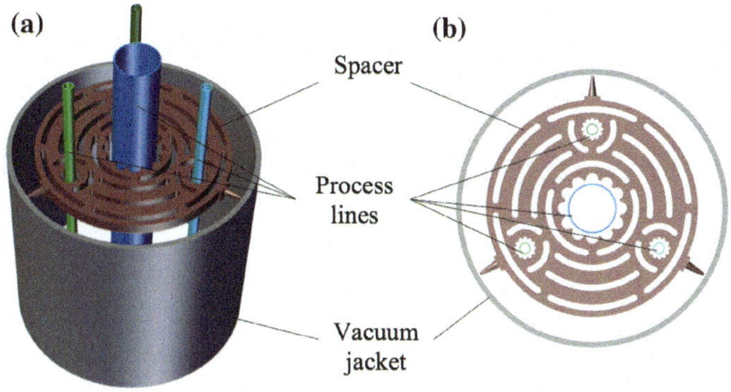


**Fig. 9.11** Process line spacer: **a** isometric view, **b** top view

of two plates connected to each other with three stiffening rods. This solution protects the plates against tilting and blocking process line movements. The dead weight is transferred to the external envelope via a set of sliding batons. The batons have spherical ends to reduce the area of contact with the vacuum jacket internal surface. In order to minimize friction the batons can be replaced with wheels, rollers or balls.

Spacers are supporting structures of a special type. Their main function is to keep the lines in proper locations in cryoline cross sections in order to avoid unwanted thermal contacts. They do not bear significant loads from process lines but only limit the movements of the lines in their radial directions. Figure 9.11 shows an example spacer design. The sizes of the holes for process lines are distinctively larger then the line sizes. Since the spacer does not bear any significant mechanical load it can be made of a material of low thermal conductivity (that usually implies lower mechanical strength) and can have a form that elongates the heat transfer paths, which additionally lessens its mechanical properties but does not add any substantial heat loads to the cold lines.

The family of cryoline external envelope supports also includes fixed supports, sliding supports and spacers. These supports carry the entire cryoline weight and transfer all forces to the surrounding structures. The forces, apart of dynamic and static forces from process lines, can include also forces resulting from thermal and pressure loads acting on the external envelope itself. The vacuum jacket (external envelope) can be exposed to a significant variation of ambient temperature and it can also get cold locally due to a cryogen leak from a cold process line in case of a failure mode. Some additional pressure forces can appear if the set pressure of the external envelope safety devices significantly exceeds the atmospheric pressure. The total forces that the fixed supports of the external envelope have to bear can reach the value of 10 kN. Figure 9.12 depicts example designs of external envelope fixed and sliding supports. The fixed support has a hoop that increases the mechanical strength of the support. The hoop is made of steel plates and is welded to the external envelope through two steel patches and to the support base. The base of the support is to be screwed to a dedicated component of the surrounding structures.

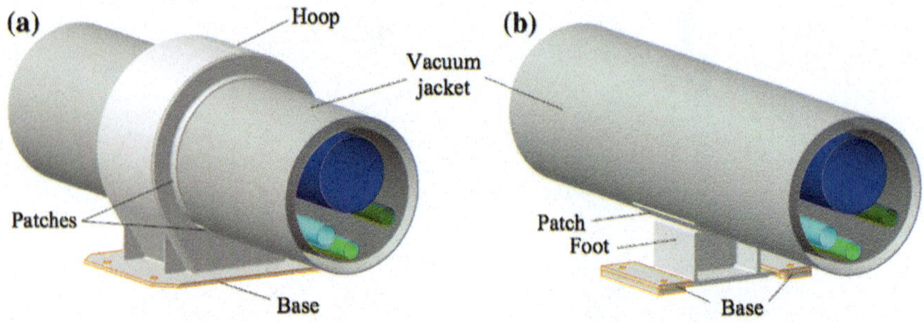


**Fig. 9.12** External envelope supporting elements: **a** fixed support and **b** sliding support

The sliding support shown in Fig. 9.12b is designed for allowing some axial movements of the cryoline. The support foot is welded to the vacuum jacket via a steel patch. The support base is to be screwed to the surrounding structure but not fixed to the support feet at all. The base blocks the lateral movements and rotations of the support foot. Usually the expected movements of the sliding support feet are not higher then a couple of centimeters.

## 9.5 Thermal Contraction Compensation

All the components of cryogenic transfer lines tend to contract due to temperature variations. This pertains not only to components which are at cryogenic temperatures (cold process lines and thermal shields) but also their vacuum jackets, since they can be exposed to a variation of the ambient temperature as well as to a significant temperature drop during a failure mode of a cold process line break. Therefore, both the process lines and external envelope must be protected against excessive stresses and forces that can result from thermal contraction [11]. Cryoline thermal contraction compensation systems can employ axial expansion joints, metallic flexible hoses and natural compensation loops.

Natural compensation loops require a lot of space, so they are used only if the cryoline routing is adequately complex and have a number of elbow and angular modules located close to each other. Taking this solution requires detailed and complex thermo-mechanical analyses of cryoline behavior in the conditions of all possible operation and safety modes. Metallic flexible hoses are usually used in elbow or angular modules to compensate some rather small dislocations of process lines. There are usually two hoses installed in the bent section of a process line. Then the contractions of the line sections are absorbed by the lateral deformations of the hoses.

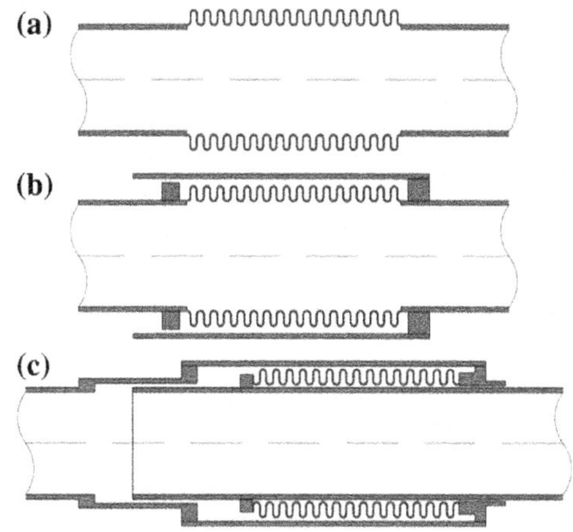


**Fig. 9.13** Typical axial expansion joints used for cryoline process lines: **a** with bellows subjected to internal pressure, **b** with bellows subjected to internal pressure and equipped with a guiding sleeve **c** with bellows subjected to external pressure

Axial expansion joints seems to be the most often used thermal-contraction compensation components, even though they are considered as the main items that can significantly lower the reliability of transfer lines [7, 25, 26]. The joints consist of bellows made of one to a few plies whose thickness has to be as small as a fraction of millimeter (usually from 0.2 to 0.5 mm). Since these very thin elements are subjected to significantly high and cyclical thermo-mechanical loads, there is a risk not only of leaks but also of loosing the mechanical stability of process lines.

Figure 9.13 depicts three typical types of axial expansion joints installed in the process pipes of cryogenic transfer lines . The first one (Fig. 9.13a) is composed of two pipe sections and bellows. It is the simplest joint but also very sensitive to become unstable. If the pipe with such a joint is not properly guided and anchored or the expansion joint has very low stiffness the bellows or even whole pipe segment can buckle [27]. This can of course lead to unwanted thermal contacts increasing heat loads and what is more dangerous, to leaks in the cold process lines.

Axial expansion joints equipped with special guiding sleeves, as shown schematically in Fig. 9.13b, can mitigate the risk of buckling to a certain extend, if the risk of local instability comes from insufficient stiffness of the bellows. However, this configuration also requires a precise arrangement of anchors and guides.

Joints with bellows subjected to external pressure (see Fig. 9.13c) are the most stable from a mechanical point of view. When such joints are pressurized they actually become less sensitive to the sources of local instabilities. On the other hand their production is significantly much more expensive.

## 9.6 Materials

Materials used for manufacturing cryogenic transfer lines are almost always the same as for other type of cryostats. Process pipes and their strong fixed supports as well as vacuum barriers are made of austenitic stainless steels adequate for temperatures which lines are going to be cooled down to. External envelopes can be manufactured of some low carbon steels, as long as the lowest expected temperature of the vacuum jacket will not cause any brittle fractures. The ductile-brittle transition of low carbon steels ends in a temperature of −50 °C, so the application of low carbon steels is not recommended if the vacuum jacket temperature can drop below this value. Then external envelope should also be made of austenitic stainless steel.

Since cryogenic transfer lines are dedicated for transferring fluids at pressure usually higher than 0.5 bar gauge, their process lines are considered as pressure equipment and have to be designed and constructed in respect to dedicated regulations and standards applicable in the country where the lines are to be installed and operated. Example documents are the ASME B31 of the USA and the European metallic industrial piping code EN 13480, harmonized with the Pressure Equipment Directive 97/23/EC (PED) of the European Union. All these standards include a list of materials that are applicable for use at given cryogenic temperatures. Cold process lines are usually made of the standard Ni-Cr low-carbon stainless steel 1.4306 (304L), due to its significantly low price and applicability in a whole cryogenic temperature range. For external envelope usually the stainless steel 1.4301 (304) is used, which is cheaper than grade 1.4306 but applicable down to 77 K only.

Cryoline thermal shield components are usually made of some aluminum alloys of 6000 series. These alloys have good thermal conductivity, low outgassing rate and the surfaces of their commercial sheets, pipes and profiles have significantly low emissivity coefficient. They are also easily machined and highly weldable by using gas tungsten arc welding technique. AA6061, which is a good medium-strength all-purpose aluminum alloy, is used for thermal shields made of pipes of sheets, whilst AA6063 alloy is used for thermal shield made of extruded shapes.

Sliding supports and spacers of the process lines are usually made of fiber-epoxy composites, such as NEMA grades G-10, G-10-FR4 which is fire retardant, or G-10CR that is characterized by much smaller variability of their mechanical and thermal properties at cryogenic temperatures. In case of cryolines designed for long-time operation in radiation environment their sliding supports can be made ofG-11 or G-11CR composites, which are less sensitive to radiation and thus retain their properties longer.

## 9.7 Manufacturing and Installation

Cryoline modules are assembled from the inner to the outer parts. Their fabrication usually consists of the following works:

1. Welding of process line sections and external envelope sections (including all required examinations of welds, such as visual examination, radiographic or ultrasonic tests, etc.),
2. Cleaning the external and internal surfaces of the prepared sections,
3. Leak tightness tests of process line and external envelope sections,
4. Winding MLI on the process lines,
5. Installation of process line fixed and sliding supports,
6. Installation of thermal shield elements,
7. Winding MLI on the thermal shield,
8. Inserting the process line sections with their thermal shield into the external envelope section,
9. Connection of the process line fixed support to the external envelope section,
10. Preparation of the cryoline module for transportation.

Cryoline interconnections are also assembled from their inner parts to outer ones, and their assembling requires the following activities:

1. Sliding the interconnection sleeve over the end of one of the cryoline modules,
2. Welding the compensators or suitable pipe sections to the ends of the process line sections,
3. Testing the process line welds (visual examination, radiographic or ultrasonic tests, leak tightness tests, pressure test, if needed, etc.),
4. Winding MLI on the process lines,
5. Installation of thermal shield segments,
6. Winding MLI on the thermal shield,
7. Sliding the interconnection sleeve over the interconnection and welding it to the welding rings of the cryoline modules,
8. Testing the sleeve welds (visual examination, leak tightness test, etc.).

Typically the installation of a cryogenic transfer line requires assembling of a number of interconnections. In order to minimize the time and cost of the installation works the leak tightness test of the process pipe welds are performed in one turn for all the interconnections. The leak tightness test of the sleeve welds are also performed

## 9.8 Case Study: XFEL/AMTF Cryogenic Transfer Line

A good example of a multichannel cryogenic transfer line is the XFEL/AMTF cryoline. This line is used for transferring cryogenic cooling power from the HERA refrigerator to the Accelerator Module Test Facility (AMTF) in the national research center of the Deutsches Electronen-Synchrotron (DESY). Figure 9.14 shows schematically the run of the cryoline in the DESY site. The AMTF is intended for testing the superconducting cavities and cryomodules of the European X-ray Free Electron Laser (XFEL) [28]. The test facility requires for its continuous

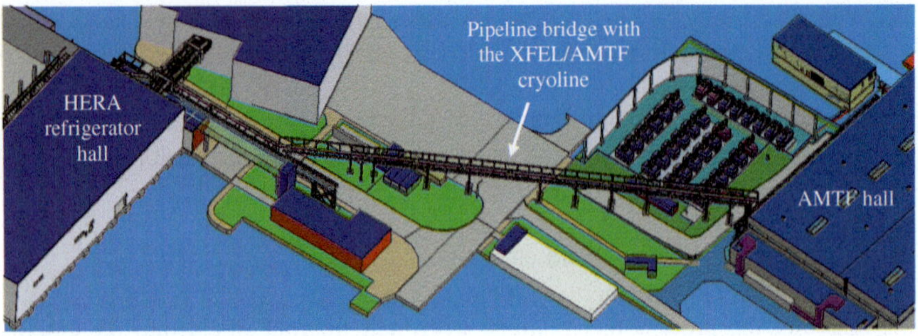


**Fig. 9.14** Schematic run of the XFEL/AMTF cryogenic transfer line (*courtesy* DESY)

operation the cooling capacities of 0.8 kW at 2.0 K, 0.5 kW at 4.5 K and 3 kW at 40/80 K [20]. Since the 2 K helium is produced by isenthalpic expansion of the 4.5 K helium (previously subcooled to 2.2 K in counter flow heat exchangers) in the facility cryostats and test boxes, the XFEL/AMTF cryogenic line houses only 4.5 and 40/80 K circuits.

The XFEL/AMTF cryoline project was executed by Wrocław Technology Park AB under the Polish in-kind contribution to the XFEL project and coordinated by National Centre for Nuclear Research, Poland [29]. All the requirements for the cryoline were provided to the Polish partners in the comprehensive specification prepared by DESY [30]. Based on this specification, Wrocław University of Technology developed the detailed design of the line and supervised production and participated in the acceptance tests. The line was manufactured and installed in the DESY site by Kriosystem Ltd [20]. The execution of the cryoline project went through the following phases:

Phase 0. Specification of requirements,
Phase 1. Cryoline design,
Phase 2. Cryoline module productions,
Phase 3. Transportation of the cryoline modules to the DESY site,
Phase 4. Installation of the cryoline modules in the site,
Phase 5. Cryoline commissioning.

The first cool down of the cryoline to the nominal temperatures was successfully curried out in August 2012 and since December 2012 the line is in continuous operation.

### *9.8.1 Technical Requirements*

All the essential requirements for the XFEL/AMTF cryogenic transfer line were prepared on the basis of DESY experience with cryogenic transfer lines for HERA and FLASH. The specification defined the run of the cryoline and its interfaces to

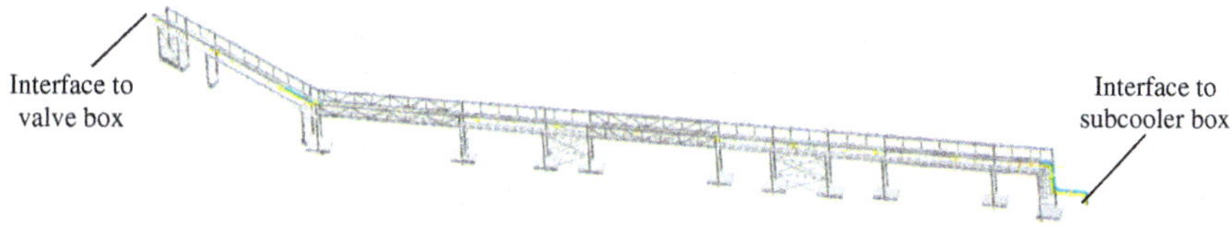


**Fig. 9.15** Routing of the XFEL/AMTF cryogenic transfer line on the dedicated pipeline bridge (*courtesy* DESY)

the adjoining components [30]. The detailed routing of the line is shown in Fig. 9.15. The line runs from a valve box located between the HERA refrigerator hall and the HERA West building on a pipeline bridge in a height of 8 to 10 m above the ground level. At the AMTF hall the line goes 8.4 m down in two steps and ends at a subcooler box located inside the hall.

The line had to be designed, manufactured and installed in conformance with Pressure Equipment Directive 97/23/EC. In order to reach the conformity with the directive, the AD2000 Pressure Vessel Code [31] had to be applied. The specification also described all the essential requirements for mechanical design, tests to be carried out at manufacture workshop and DESY site and guarantee data. The defined sizes and operating conditions of the process lines, thermal shield and external envelope are given in Table 9.2.

The line had to be designed so as to ensure reliable and uninterruptible operation for at least 5 years and withstand 200 cool down/warm up cycles without damage and deterioration of quality. All pressure bearing components (process lines and their supports) had to be designed for the maximum pressure of 20 bar against vacuum and for the pressure of 0 bar against atmospheric pressure at all possible operating temperatures. Similarly, all the components of external envelope and their

**Table 9.2** Sizes and operating conditions of the process lines, thermal shield and vacuum jacket

| Pipe | Size | Diameter and thickness | Design pressure (bara) | Operating pressure (bara) | Operating temperature (K) |
|---|---|---|---|---|---|
| 4.5 K supply | DN50 | 60.3 mm × 2 mm | 20 | 3.5 | 4.5–6 |
| 4.5 K return | DN80 | 88.9 mm × 2.3 mm | 20 | 1.2 | 4.5 |
| 40 K supply | DN40 | 48.3 mm × 2 mm | 20 | 17 | 40 |
| 80 K return | DN40 | 48.3 mm × 2 mm | 20 | 16.7 | 80 |
| Thermal shield | NA | 300 mm × 4 mm | NA | NA | 80 |
| Vacuum jacket | DN400 | 406.4 mm × 4.78 mm | 1.5 | −1 | 300 |

supports had to be designed for the internal pressure from 0 bar absolute to 1.5 bar absolute against the atmospheric pressure. In addition, the process lines had to be designed so as to withstand the maximal possible temperature difference between the inlet and outlet. It should also allow for independent cool-down/warm-up cycles of the different process circuits as well as for rapid cool-downs or warm-ups of any circuit.

Requirements on the tightness of the process lines and external envelope defined the values of allowable single and integral leak rates into the isolation vacuum. The single leak rate from any component of process pipes and external envelope such as welds, bellows and corrugated hoses should not be higher than $1 \times 10^{-9}$ mbar dm$^3$/s at the design pressure and both room and operating temperatures, whilst the integral leak rate to the insulation vacuum should not exceed $1 \times 10^{-8}$ mbar dm$^3$/s.

Maximum allowable heat loads to the thermal shield and 4.5 K process lines should not be higher than 1.5 and 0.15 W per meter, respectively. The cryoline should be aligned so as the connections to the valve and subcooler boxes are achieved within the limits of lateral forces for the process pipes and the vacuum shells of the connection ports of the boxes.

The design of the line also had to take into account the variations of ambient temperature in the range from 260 to 310 K and all possible thermal reactions resulting from the expansions and contractions of the bridge components.

## 9.8.2 Design

Due to a significantly high distance between a potential manufacturer workshop and the DESY site it was decided that the sizes of the cryoline modules should not exceed the dimensions of a standard European semitrailer (13.62 m × 2.75 m × 2.48 m). So the maximum length and width of the modules were taken as 13 and 2.5 m, respectively [32]. The modularization of the cryogenic transfer line is shown in Fig. 9.16. The line is composed of 11 straight modules, 3 elbow modules and 1 angular module with an angle of 150.5° [20]. The cross section of the line is shown in Fig. 9.8.

The interconnections between the modules were designed as typical sleeve-shaped interconnections holding the components of the process line thermal compensation systems. Figure 9.17 shows the schematic layout of the process line

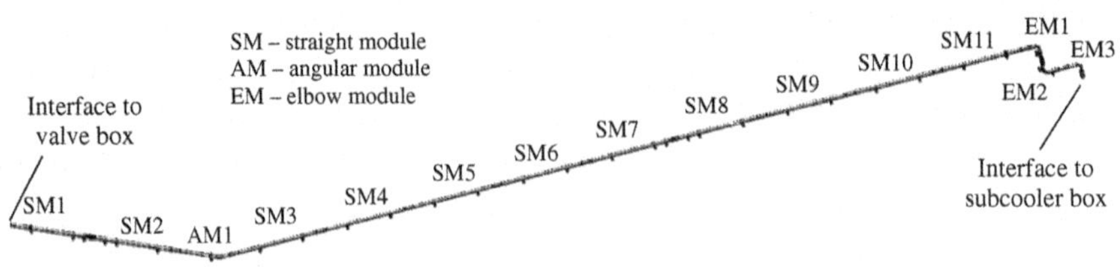


**Fig. 9.16** Modularization of the XFEL/AMTF cryoline

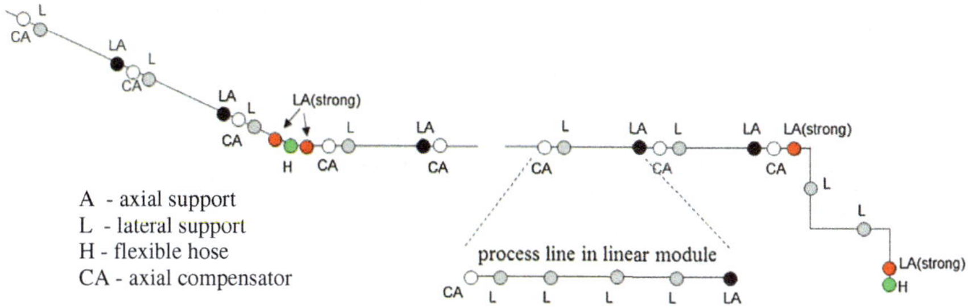


**Fig. 9.17** Schematic layout of the process line support and compensation systems

supporting and thermal compensation systems [33]. These systems include axial compensators, flexible hoses, fixed supports and sliding supports. The sliding supports block lateral movements, whilst the fixed supports block both lateral and axial movements. Due to accumulation of the pressure forces at the extremities of the cryoline straight sections, each line requires only four strong fixed supports and 11 fixed supports of considerably smaller mechanical strength. The thermal contraction of each line is compensated by a set of 13 axial expansion joints and 2 flexible hoses. Since the three elbows of the process lines are very close to each other this section works as a natural compensation loop and there is no need for any expansion joints.

The external envelope acts on the bridge not only with its own pressure, thermal and dead weight loads but it also transfers the loads from internal process lines. The layout of its supporting and thermal compensation systems is schematically shown in Fig. 9.18. This supporting system consists of two fixed supports which block vertical, lateral and axial movements (VLA) in respect to the piping bridge. There are also 21 sliding supports allowing for axial movements only (VL) and 2 sliding supports blocking the line movements only in vertical directions (V). The thermal expansions or shrinkages of the envelope are compensated by three expansion joints. Each of the two long straight sections are equipped with axial compensators (CA). The thermal deformations in the section of the three elbow modules are compensated with a lateral expansion joint (CL), which is built in the interconnection between module E90-3 and the Subcooler Box.

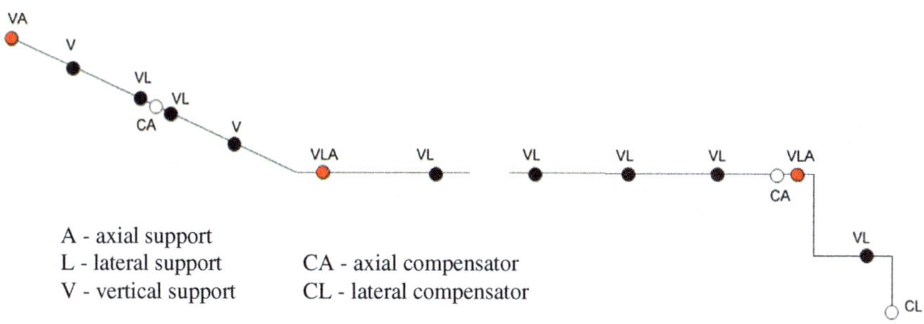


**Fig. 9.18** Schematic layout of the external envelope support and compensation systems

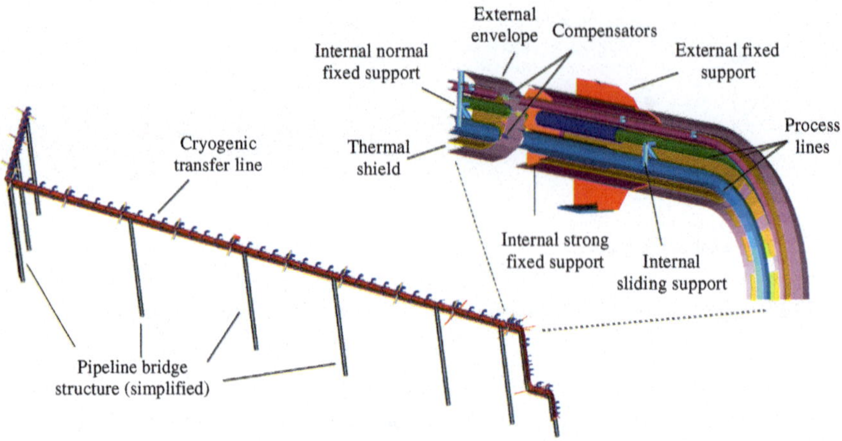


**Fig. 9.19** FEM model of the XFEL/AMTF cryogenic transfer line (*courtesy* Wrocław University of Technology)

The thermal and mechanical behavior of the cryogenic transfer line with the proposed support and compensation systems was analyzed numerically. For this purpose a complex Finite Element Method model of the cryoline and its pipeline bridge was built as shown in Fig. 9.19 [34]. The model consisted of 107,000 elements and included all the kinematic joints of the external envelope, thermal shield and process line supports. It used shell elements for process lines, thermal shield and external envelope, beam elements for sliding supports and pipeline bridge structure, spring elements for all the compensators and finally membrane elements for elastic hoses. The model took into account the variation of the material properties in the specified temperature ranges. It was assumed that the process lines are made of stainless steel of grade 1.4306 or 1.4541 ($A_5 = 40$ %, $R_{p1,0} = 220$ MPa and $R_m = 520$ MPa) and the external envelope is made of SS1.4301 ($A_5 = 43$ %, $R_{p1,0} = 235$ MPa and $R_m = 540$ MPa).

The thermo-mechanical strength analyses of the entire cryoline were carried out for five cases [33]:

Case 1: Design conditions (process lines and at their design pressure and operating temperatures, thermal shield at 80 K and external envelope under vacuum and at 285 K, which is an assumed assembling temperature),

Case 2: Pressure test conditions (process lines at their test pressure (28.6 bara) and 260 K, which is the lowest possible temperature),

Case 3: Failure mode I conditions (the external envelope of SM2 and AM1 at 160 K and the other cryoline components at the design conditions),

Case 4: Failure mode II conditions (the external envelope of SM10 and SM11 at 160 K and the other cryoline components at the design conditions),

Case 5: Failure mode III conditions (the external envelope of EM1, EM2 and EM3 at 160 K and the other cryoline components at the design conditions).

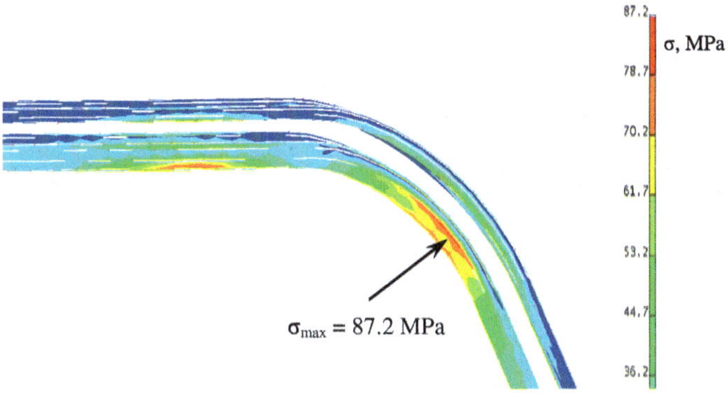


**Fig. 9.20** Von Mises stress distribution on the EM1 process lines (*courtesy* Wrocław University of Technology)

The thermo-mechanical strength analyses resulted in the distributions of von Mises stresses and deformations as well as the forces acting on the supports. The obtained maximum stresses on the external envelope and process lines were equal to 80.4 and 87.2 MPa, respectively. Fig. 9.20 shows an example picture of the von Mises stress distribution on the process line sections. The maximum allowable stresses, which take into account a safety factor of 1.5, are equal to 156.7 MPa for the external envelope and 146.7 MPa for the process lines. So, the numerical modeling showed that the maximum stresses did not exceed 75 % of the allowable stress values and therefore, according to [35], both the vacuum jacket and process lines are in stress category II. It means that the lowest temperatures of the external envelope and process lines can reach −255 and −273 °C, respectively, what approves the selection of the materials.

The obtained values of deformations of process lines and forces acting on all the external and internal supports were used as input for designing these supports and selecting adequate expansion joints. The 3D models of the designed process lines supports are shown in Fig. 9.21. The process line strong fixed support is made of stainless steel of grade 1.4306 or 1.4541, whilst the normal fixed support and sliding supports are made mainly of G10 (marked in green). The double sliding support is made of two single supports fixed together with three metal rods and 3 G10 tubes. Such a support, apart of blocking lateral movements of the process lines in respect to the thermal shield, protects also the bunch of these lines against twisting. So each straight section is equipped at one end with a normal fixed support and at the other with a double sliding support.

The G-10 plate of the normal fixed supports is fixed to the radiation shield, which is made of ∅300 AA6060 pipe sections. The sliding supports are mechanically fixed only to the 4.5 K return line (DN 80). The other pipes are not connected to the support and are allowed to move axially, so all the process pipes can contract and expand independently. The sliding support plates are equipped with stainless steel rollers what allow them to move axially inside the radiation shield.

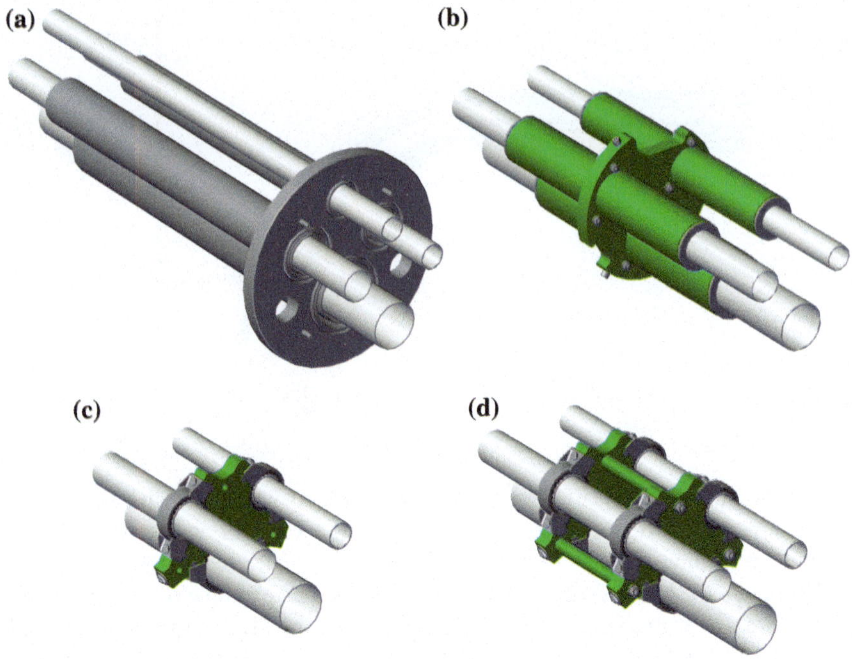


**Fig. 9.21** 3D models of the process line strong fixed support (**a**), normal fixed support (**b**), single sliding support (**c**) and doubled sliding support (**d**) (*courtesy* Wrocław University of Technology)

## *9.8.3 Manufacturing the Cryoline Modules*

All the 11 straight, 3 elbow and 1 angular modules of the XFEL/AMTF cryogenic transfer line were manufactured by Kriosystem Ltd. in the period of 10 months. The assembly works of the straight modules contained the following set of technical operations:

1. Welding of process line sections and external envelope sections,
2. Test of the process line and external envelope sections,
3. Manufacturing components of the thermal shield,
4. Cleaning the surfaces of the process lines, external envelope and thermal shield,
5. Winding MLI on the 4.5 K and 40 K process lines,
6. Installation of the supports on the 4.5 K and 40 K process line sections,
7. Assembling the 80 K return line sections with the thermal shield components,
8. Inserting the 4.5 K and 40 K process lines into the thermal shield sections,
9. Fixation of thermal links to the thermal shield and the 80 K return line,
10. Winding MLI on the thermal shield sections,
11. Installation of the sliding supports to thermal shield sections,
12. Inserting the thermal shield sections in the external envelope sections,
13. Connection of the process line fixed supports to the external envelope sections.

**Fig. 9.22** Sliding supports on the XATL1 straight module process lines (*courtesy* Kriosystem Ltd.)

As soon as each of the cryoline modules was assembled it was packed and prepared for its transportation to the DESY site. Figures 9.22, 9.23, 9.24 and 9.25 show the photos taken during the production of a straight module. Figure 9.22 shows process lines subassembly at technical operation 6 (double sliding support in the foreground). Figures 9.23 and 9.24 show one of the straight module subassemblies under technical operations 9 and 10, respectively. One of the fully assembled straight modules is shown in Fig. 9.25. Its process lines are ended with axial expansion joints enclosed in guiding sleeves (as schematically presented in Figure 9.13b).

## 9.8.4 Installation

The modules of the XFEL/AMTF cryogenic transfer line just after their manufacturing were packed and later transported to the DESY site. The installation work included the following steps:

**Fig. 9.23** Aluminium thermal shield of the XATL1 straight module with its copper thermal links to the thermal shield return line (*courtesy* Kriosystem Ltd.)

**Fig. 9.24** MLI radiation foils wrapped on the XATL1 straight module thermal shield (*courtesy* Kriosystem Ltd.)

**Fig. 9.25** Straight module assembled (*courtesy* Kriosystem Ltd.)

1. Lifting the cryoline modules on the pipeline bridge,
2. Positioning of the cryoline modules,
3. Fixing the external supports to the piping bridge,
4. Welding the process line sections of all the cryoline modules and connecting them to the process lines of the valve box and subcooler box,
5. Testing the welding seams of the process lines put during the installation,
6. Executing the pressure test of the entire process lines,
7. Wrapping MLI on the process line interconnection,
8. Installation of the thermal shield components in the cryoline interconnections,
9. Wrapping MLI on the thermal shield interconnections,
10. Closing the external envelope sleeves and welding them to the vacuum vessel rings,
11. Testing the welding seams of the vacuum vessel put during the installation,
12. Performing the leak tightness test of the entire cryogenic line.

During the installation some inner parts of the cryoline could be exposed to weather conditions. Therefore all the interconnections located on the bridge were protected with special tents that were conditioned in order to keep the inner parts dry and clean. Figure 9.26 shows a cryoline straight module being lifted on the pipeline bridge (step 1) and Fig. 9.27 shows the tents. The installation of the interconnection between the cryoline and subcooler box (step 4) is presented in Fig. 9.28. Here, the process lines are being connected with flexible hoses (protected with braids) and the lateral expansion joint is suspended above the interconnection. When all the internal parts of the interconnection were installed, the lateral compensator was moved down and welded to the dedicated welding rings on the vacuum jacket of the cryoline and the subcooler. Figure 9.29 presents the XFEL/AMTF cryogenic transfer line after its installation. In the foreground there is the

**Fig. 9.26** Installation of one of the XFEL/AMTF cryoline straight modules on the pipeline bridge (*courtesy* Kriosystem Ltd.)

**Fig. 9.27** The XFEL/AMTF cryoline interconnections on the pipeline bridge protected with tents (*courtesy* Kriosystem Ltd.)

angular module with its strong fixed support. The visible stubs on the straight section, which are covered with yellow lids, are the safety devices of the external envelope. In case of some failure modes of process line rupture these safety devices will discharge the helium from the isolation vacuum space to the surroundings and protect the external envelope against overpressurization.

**Fig. 9.28** Connecting the XFEL/AMTF cryoline to the cold terminal of the subcooler box (*courtesy* Kriosystem Ltd.)

**Fig. 9.29** The XFEL/AMTF cryogenic transfer line installed on the pipe bridge (*courtesy* Kriosystem Ltd.)

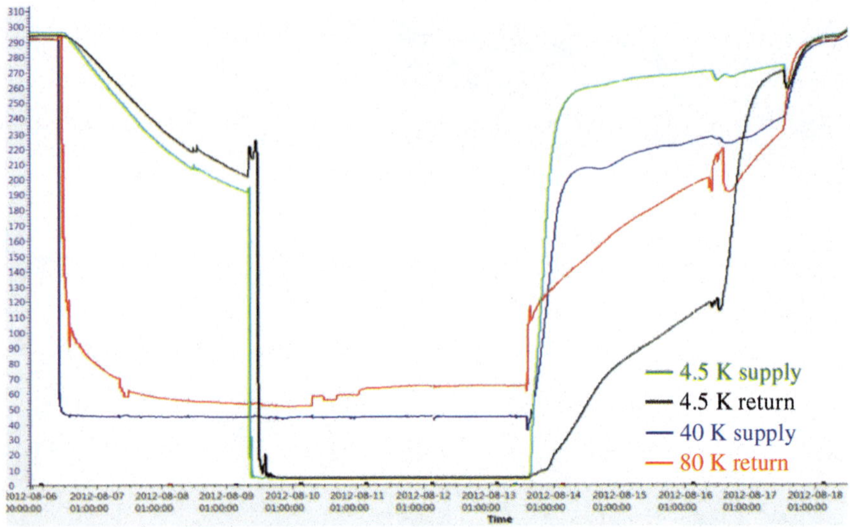


**Fig. 9.30** The evolution of temperatures in the valve box process lines measured during the commissioning of the XFEL/AMTF cryogenic transfer line (*courtesy* DESY)

### *9.8.5 Commissioning and Performance*

The preliminary acceptance test of the XFEL/AMTF cryogenic transfer line was carried out by DESY between 06.08.2012 and 18.08.2012 together with the cold performance tests of the subcooler box [28]. The measured temperatures in the valve box process lines are shown in Fig. 9.30. During the test the cryoline was rapidly cooled down close to the nominal operating conditions for a period of four days and then abruptly warmed up to ambient temperature. The data collected from the commissioning allowed for checking the thermodynamic efficiency of the cryoline by applying the Second Law analysis and Gouya-Stodola theorem. The performed entropy generation analysis showed that the thermodynamic efficiency of the cryoline was only 3 % different from the specified parameters [36]. In December 2012 the line was cooled down again and since then it is in continuous operation and serves the AMTF facility for the tests of the XFEL linac cryomodules and their cavities [28].

## References


1. J.G. Weisend II, Transfer lines, defining cryogenics, cold facts. Winter 2011 issue (2011)
2. N. Dittmar, C. Haberstroh, U. Hesse, M. Krzyzowski, Characterisation and optimization of flexible transfer lines for liquid helium. part I: experimental results. Cryogenics **75**, 6–12 (2016)

3. K. Kawano, K. Hamada, T. Kato, T. Honda, K. Nishida, K. Matsui, T. Hiyama, K. Ohtsu, S. Sekiguchi, H. Tsujji, M. Ando, T. Hiyama, K. Ichige, Design and construction of long cryogenic piping lines, in *Proceedings of the Sixteenth International Cryogenic Engineering Conference* (Oxford, 1997), pp. 493–496
4. McIntosh, Vacuum-jacketed transfer line design, section 5-6, in *Handbook of Cryogenic Engineering*, ed. by J.G. Weisend II (Taylor & Francis Ltd, 1998), pp. 269–278
5. W. Erdt, G. Riddone, R. Trant, The cryogenic distribution line for the LHC: functional specification and conceptual design, in *Advances in Cryogenic Engineering, Transactions of the Cryogenic Engineering Conference—CEC*, vol. 45 (2000), pp 1387–1394
6. B. Barbier, A. Bouillot, K. Brodzinski, M. Dupont, M. Fathallah, J-L. Fournel, E. Gitton, S. Junker, H. Moussavi, C. Parente, G. Riddone, Manufacturing and installation of the compound cryogenic distribution line for the large hadron collider, in *Proceedings Twenty First International Cryogenic Engineering Conference* (Praha, Czech Republic, 2006), pp. 31–34
7. M.A. Iarocci, D. Brown, J. Sondericker, K.C. Wu, J. Benson, Y. Farah, C. Lac, A. Morgillo, A. Nicoletti, E. Quimby, J. Rank, M. Rehak, A. Werner, RHIC cryogenics. Nucl. Instrum. Methods Phys. Res. A **499**, 264–279 (2003)
8. V. Kalinin, E. Tada, F. Millet, N. Shatil, ITER cryogenic system. Fusion Eng. Des. **81**(23–24), 2589–2592 (2006)
9. B. Sarkar, N. Shah, H. Vaghela, R. Bhattacharya, K. Choukekar, P. Patel, M. Chalifour, Value engineering in system of cryoline and cryodistribution for ITER: in-kind contribution from india. Paper presented in CEC-ICMC2015. Tucson, USA, 2015
10. J. Fydrych, P. Arnold, W. Hees, P. Tereszkowski, X.L. Wang, J.G. Weisend II, Cryogenic distribution system for the ESS superconducting proton linac. Phys. Proc. **67**, 828–833 (2015)
11. J. Fydrych, M. Chorowski, J. Polinski, J. Skrzypacz. Design methodology of long complex helium cryogenic transfer lines, in *Advances in Cryogenic Engineering, Transactions of the Cryogenic Engineering Conference—CEC*, vol. 55 (2010) pp. 1103–1110
12. K. Hosoyama, K. Hara, A. Kabe, Y. Kojima, T. Ogitsu, Y. Sakamoto, S. Kawamura, K. Matsumoto, Cryogenic system For TRISRAN superconducting RF cavity. Fusion Eng. Des. **20**, 491–498 (1993)
13. G. Bon Mardion, B. Gravil, B. Pager, The tore supra, flexible and electrically insulating, cryogenic connection line. Cryogenics **32**(Supplement 1), 142–145 (1992)
14. G. Bon Mardion, Systeme cryo, les canalisations cryogeniques. Association Euroatom CEA Note TS-71-93-02, Cadarache (1993), p. 30
15. B. Sarkar, S. Badgujar, H. Vaghela, N. Shah, R. Bhattacharya, C. Chakrapani, Design, analysis and test concept for prototype cryoline of ITER, in *Advances in Cryogenic Engineering: Transactions of the Cryogenic Engineering Conference—CEC*, vol. 53 (1998), pp. 1716–1723
16. E. Pyata, L. Belova, T. Boeckmann, M. Kholopov, V. Konstantinov, V. Kulikov, D. Sellmann, A. Zhirnov, N. Zolotukhina, XFEL injector-1 cryogenic equipment. Phys. Proc. **67**, 868–873 (2015)
17. V. Ganni, K. Dixon, N. Laverdure, P. Knudsen, D. Arenius, M. Barrios, S. Jones, M. Johnson, F. Casagrande, FRIB cryogenic distribution system, in *Advances in Cryogenic Engineering: Transactions of the Cryogenic Engineering Conference—CEC*, vol. 59 (2014), pp. 880–886
18. V. Ganni, K. Dixon, N. Laverdure, S. Yang, T. Nellis, S. Jones, F. Casagrande, FRIB cryogenic distribution system and status. Paper presented in CEC-ICMC2015, Tucson, USA, 2015
19. Liquid helium transport to the hera tunnel, the transfer line in DESY webpage: http://www-mks2.desy.de/content/e3740/e4847/e5250/e6587/e6593/e6628/e6743/index_ger.html
20. J. Fydrych, M. Chorowski, P. Duda, P. Grzegory, A. Iluk, K. Malcher, G. Michalski, E. Rusinski, G. Strychalski, Design, manufacturing and assembly of the cryogenic transfer line for XFEL/AMTF, in *Proceedings of the 12th IIR International Conference on Cryogenics, Cryogenics 2012* (Dresden, Germany, 2012), pp. 28–33

21. E.C. Quimby, C.M. Lac, M. Iarocci, R. Sallash, A. Varghese, VJR/VJRR design, construction, installation, and performance, in *Advances in Cryogenic Engineering: Transactions of the Cryogenic Engineering Conference—CEC*, vol. 43 (1998), pp. 531–540
22. The fermilab tevatron cryogenic cooling system. Fermi National Accelerator Laboratory, United States Department of Energy, Batavia, AMSE getmedia (1993)
23. K. Hosoyama, K. Hara, A. Kabe, Y. Kojima, Y. Morita, H. Nakai, T. Fujita, T. Kanekiyo, K. Matsumoto, Development of a high performance transfer line system. Adv. Cryog. Eng. **45**, 1395–1402 (2000)
24. J. Yoshida, K. Hosoyama, H. Nakai, K. Hara, Y. Kojima, K. Nakanishi, M. Noguchi, T. Ichitani, S. Kaneda, S. Sakuma, K. Suzuki, T. Kanekiyo, Development of STF cryogenic system in KEK, in *Proceedings of PAC07, Albuquerque* (New Mexico, USA, 2007), pp. 2701–2703
25. K. Brodzinski, P. Cruikshank, J.L. Fournel, L. Tavian, N. Veillet, Failure mechanism and consolidation of the compensation bellows of the LHC cryogenic distribution line. Physics Procedia **67**, 129–134 (2015)
26. A. Thakkar, M.I. Vyas, Design & analysis of bellows free cryogenic transfer line. Presented at the international conference on current trends in technology 'NUiCONE-2011', Institute of Technology, Nirma University, Ahmedabad, 2011
27. B. Skoczen, Stability, Fatigue and optimization of thin-walled structures under cryogenic conditions—application in the structural design of colliders and cryogenic transfer lines. CERN Yellow Report CERN-2001-001, p. 121 (2001)
28. B. Petersen, Full cryogenic operation of XFEL accelerator module test facility—the first year, in *Oral Contribution to the 6th International Workshop on Cryogenics Operation presented*, 11 Nov. 2014. Electronic copy available at https://eventbooking.stfc.ac.uk/news-events/6th-international-workshop-on-cryogenics-operations-218?agenda=1/
29. E. Pławski, J. Sekutowicz, W. Grabowski, K. Kosiński, J. Lorkiewicz, M. Wojciechowski, Z. Gołębiewski, K. Meissner, G. Wrochna, M. Duda, M. Jeżabek, K. Kasprzak, A. Kotarba, K. Krzysik, M. Stodulski, J. Świerblewski, M. Wiencek, M. Chorowski, E. Rusiński, J. Fydrych, A. Iluk, K. Malcher, J. Poliński, P. Duda, J. Głowinkowski, P. Wilk, M. Winkowski, P. Grzegory, G. Michalski, Polish in-kind contribution to European X-ray free electron laser (XFEL)-status in summer 2014, chapter 3, in *Vacuum Technique & Technology* (Monographs of Tele & Radio Research Institute, Poland, 2014), pp. 33–43
30. D. Sellmann, Y. Bozhko, B. Petersen, XFEL project. Technical Specification for the Cryogenic Transferline XATL1, Deutsches Elektronen-Synchrotron, MKS-S0008.0 (2009), p. 34
31. AD2000-Merkblatt B0, Design of pressure vessels, ICS 23.020.30, Nov 2008 edition
32. J. Fydrych, Functional description of the cryogenic transfer line XATL1, Technical document WUT_ XATL1-TD_001-0111 Ver. 1, Wrocław University of Technology (2011), pp. 1–4
33. A. Iluk, Design calculation of the cryogenic transfer line XATL1, Technical document WUT_ XATL1-TD_002-0111 Ver. 3, Wrocław University of Technology (2011), pp. 1–32
34. E. Rusinski, M. Chorowski, A. Iluk, J. Fydrych, K. Malcher, Selected aspects related to the calculations and design of a cryogenic transfer line. Arch. Civ. Mech. Eng. **14**, 231–241 (2014)
35. AD2000-Merkblatt W10, Materials for pressure vessels, materials for low temperatures, ferrous materials, ICS 23.020.30, Nov 2007 edition
36. P. Duda, M. Chorowski, J. Polinskim, Design, optimization and operational parameters of multichannel cryogenic transfer line for XFEL AMTF, in *Poster Contribution 367 to the 25th International Cryogenic Engineering Conference*, 9 July 2014. Electronic copy available at https://indico.cern.ch/event/244641/session/35/contribution/367/

# Chapter 10
# Guidelines for Successful Cryostat Design

J.G. Weisend II

**Abstract** This chapter acts as a summary of the book by providing a set of guidelines for successful cryostat design.

## 10.1 Introduction

The information in previous chapters along with the extensive reference lists provides a solid overview of cryostat design. As a way of summarizing this work, listed below are some general guidelines to successful cryostat design.

## 10.2 Guidelines

1. Define and prioritize requirements first. Optimize the cryostat design based on these priorities.
2. Design in safety features from the start of the project.
3. Only use materials shown to be appropriate for cryogenic temperatures.
4. Review literature and learn from previous efforts. Take advantage of existing codes and standards if possible.
5. Use tested commercial solutions whenever possible.
6. Intercept heat at higher intermediate temperatures.
7. Allow for the effect of thermal contraction on cryostat alignment and design. Do not over constrain the movement of cryostat components as they cool.
8. Avoid feed throughs and demountable seals at cryogenic temperatures.
9. Be sure to properly heat sink temperature sensor wires to ensure reduced heat leak and an accurate reading.

J.G. Weisend II (✉)
European Spallation Source ERIC, Lund University, Lund, Sweden
e-mail: john.weisend@esss.se



J.G. Weisend II (ed.), *Cryostat Design*, International Cryogenics Monograph Series,
DOI 10.1007/978-3-319-31150-0_10

10. Install sensors such a pressure transducers and flow meters at room temperature when possible.
11. Analyze the design for possible thermoacoustic oscillations.
12. Conduct design reviews. These should include experts not directly involved in the design under review. Ideally, there should be at least reviews at the preliminary or conceptual level and again once the detailed design is complete. Safety should be a part of these reviews or separate safety reviews should be held.
13. Conduct prototype tests when required. Leave enough time in the design process to benefit from the results of such tests.
14. In cases where a large number of cryostats are to be produced, carry out series testing of the production cryostats in addition to any prototype testing. Allow sufficient resources (time, facilities, funding) in the project plan to accomplish these tests.

## 10.3 A Final Comment

It is hoped that by now the value of learning from cases studies in cryostat design has been shown. Workers in this field can benefit the broader cryogenics community by publishing the details and results of their cryostat designs. This includes what worked and perhaps more importantly what didn't work. There are ample venues for such information sharing including conferences, workshops and journals.

# Index





J.G. Weisend II (ed.), *Cryostat Design*, International Cryogenics Monograph Series,
DOI 10.1007/978-3-319-31150-0

The manufacturer's authorised representative in the EU is Springer Nature Customer Service Centre GmbH, Europaplatz 3, 69115 Heidelberg, Germany. If you have any concerns regarding our products, please contact ProductSafety@springernature.com

Printed and bound by CPI Group (UK) Ltd, Croydon, CR0 4YY
30/09/2026
02225435-0006